Vandana Sakhre, Uday Pratap Singh
Reactive Distillation

Also of interest

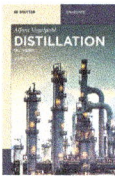

Distillation.
The Theory
Vogelpohl, 2021
ISBN 978-3-11-073972-5, e-ISBN (PDF) 978-3-11-073973-2

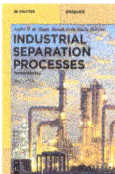

Industrial Separation Processes.
Fundamentals
De Haan, Eral, Schuur, 2020
ISBN 978-3-11-065473-8, e-ISBN (PDF) 978-3-11-065480-6

Process Technology.
An Introduction
De Haan, Padding, 2022
ISBN 978-3-11-071243-8, e-ISBN (PDF) 978-3-11-071244-5

Process Intensification.
by Reactive and Membrane-assisted Separations
Skiborowski, Górak (Eds), 2022
ISBN 978-3-11-072045-7, e-ISBN (PDF) 978-3-11-072046-4

Vandana Sakhre, Uday Pratap Singh

Reactive Distillation

Advanced Control using Neural Networks

DE GRUYTER

Authors
Dr. Vandana Sakhre
Department of Chemical Engineering
Manipal Academy of Higher Education
Dubai
United Arab Emirates

Uday Pratap Singh
Shri Mata Vaishno Devi University
Sub-Post Office
Network Centre, Katra 182320
Jammu and Kashmir
India

ISBN 978-3-11-065614-5
e-ISBN (PDF) 978-3-11-065626-8
e-ISBN (EPUB) 978-3-11-065641-1

Library of Congress Control Number: 2022931265

Bibliographic information published by the Deutsche Nationalbibliothek
The Deutsche Nationalbibliothek lists this publication in the Deutsche Nationalbibliografie;
detailed bibliographic data are available on the Internet at http://dnb.dnb.de.

© 2022 Walter de Gruyter GmbH, Berlin/Boston
Cover image: Artystarty / iStock / Getty Images Plus
Typesetting: Integra Software Services Pvt. Ltd.
Printing and binding: CPI books GmbH, Leck

www.degruyter.com

Dedicated to Parents, Husband and Son

Contents

About the author

Born in April 20, 1971, at Bhilai in Madhya Pradesh, **Dr. Vandana Sakhre** graduated in chemical engineering from the National Institute of Technology (NIT), Raipur, in 1994. She obtained her postgraduate degree in environmental science and engineering from the Indian Institute of Technology (IIT), Dhanbad, in 2002. She obtained her doctorate degree in chemical engineering in December 2016 from Rajiv Gandhi Technical University (RGPV), Bhopal (Madhya Pradesh).

Dr. Vandana Sakhre has over *23 years of experience of teaching, research, and industry*, which includes 18 years as assistant professor, 2.5 years of industrial experience, and 3 years as research fellow of CSIR Labs.

During teaching, she has developed various labs, including heat transfer, mass transfer and chemical reaction engineering and process dynamics and control labs. The advanced separation and research lab for MTech program was also developed.

Currently, *Dr. Vandana is working with Manipal Academy of Higher Education, Dubai campus*, as assistant professor-selection grade since 2017. She is handling Chemical Engineering Department as program coordinator and the coordinator of student's project and research work. She is presently working on BASF project as an external funded project, and she did few internally funded research projects. She is dedicated to teaching, learning, and research.

Dr. Vandana has worked on *artificial intelligence techniques for synthesis and control of chemical separation processes, especially reactive distillation (RD)*. She is also engaged in research on sustainability and environment management. In this book, she is giving overview and case studies on RD.

https://doi.org/10.1515/9783110656268-203

Chapter 1
Introduction

1.1 Distillation process

In the present scenario, distillation plays an important role in process industry. It is a popular separation process in chemical, petrochemical, and other industries. Its applications are in separation of crude oil into its fractions, purification of solvents, distillation of air, water, and fermented solutions, and many more. Distillation is a separation process which is based on differences in boiling points or principle of relative volatilities of the components present in the liquid mixture. Heat is given to the system so that components get separated based on their boiling points to the desired purity; hence, it is an energy-intensive separation process. The greater the difference in volatility of the component, it is easier to separate the component from their mixture. There are two types of distillation process: binary distillation and multicomponent distillation process. In binary distillation, separation of mixtures of two components takes place, while in case of multicomponent distillation, separation of more than two components takes place.

In chemical process industries, the conventional sequence of manufacturing consists of reaction followed by separation/purification. The combination of reaction and separation into one single step/unit is known as reactive distillation (RD). RD is a process in which a reaction is combined with separation of product components using distillation in a single column. Two types of RD processes are considered for the present research work, that is, continuous RD process and reactive divided wall process [1].

1.1.1 Continuous reactive distillation process

The combination of reaction and distillation performed in a single vessel is known as RD. Unlike the conventional reaction and separation process, RD is a combination of both in a single unit. Manufacturing of various chemicals like esters, ethers, cumene, and petroleum processing unit, required a reactor followed by a separator such as a distillation unit to separate the required product from other constituents based on relative volatility. There are various constraints on this type of processing like more space is required for the installation of the unit, higher cost, more energy input requirement, and reduced selectivity. Specifically, the conversion limits for reversible reactions are difficult to overcome toward highest purity of products because once the equilibrium is achieved in a system, no more reactants will be converted into products. In view of all these constraints, RD emerged as a novel technique of process

https://doi.org/10.1515/9783110656268-001

intensification in which reaction and separation of products take place simultaneously in a single column.

The RD column is divided into three sections: the upper rectifying section, middle reaction section, and bottom stripping section. The reaction takes place in the reaction section while simultaneous separation of products formed takes place in rectifying and stripping sections. The continuous removal of products from the reaction mixture makes it feasible for equilibrium-limited reactions. The simultaneous removal of products also increases the rate of reaction and prevents any undesirable side reaction between reactants and products. The RD is also preventing the formation of azeotrope. The schematic diagram of RD column is shown in Figure 1.1.

The most important benefit of RD technology is a reduction in capital investment because two process steps can be carried out on the single device. Such integration leads to lower costs in pumps, piping, and instrumentation. For exothermic reaction, the reaction heat can be used for vaporization of liquid. This leads to saving of energy costs by the reduction of reboiler heat duties [2].

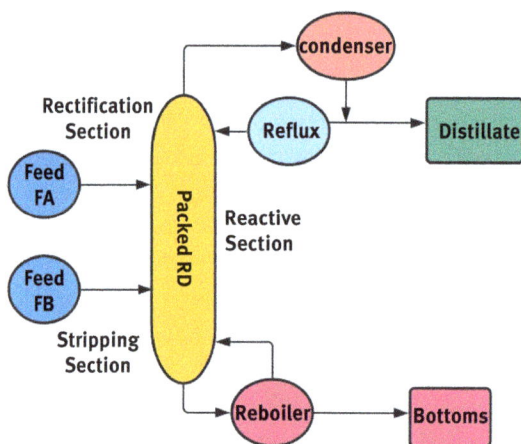

Figure 1.1: Schematic diagram of reactive distillation.

1.1.2 Reactive divided wall (RDW) distillation process

If a mixture of A, B, and C (in order of increasing boiling point) were to be separated, this could be done in one of three ways: direct, indirect, or transition split. In a direct split, component A will be the desired distillate, while a second column is required to separate B and C. In an indirect split, component C is obtained from the bottom, while A and B in the distillate require a further separation. In a transition split, A and B are from the distillate, while B and C are from the bottom. Then two more columns are needed, one for separating each of the original product streams. If these secondary columns are joined and the middle product B is taken as a side stream, and other side draws were sent back to the first column as recycles, this becomes what is known as a Petlyuk arrangement. When both columns of the Petlyuk arrangement are placed in the same shell, it becomes a dividing wall column. Combining the columns in one shell allows the removal of the condenser and reboiler from the pre-fractionators, with a consequent reduction in energy needs.

Reactive divided wall distillation column is a combination of RD and divided wall distillation column. In one part of the divided wall column, reaction takes place and the same heat of the reaction and the process is used to further separate the products in the other part of the column, thus increasing the productivity and reducing the energy requirements. Other design principles and the processes are similar as compared to RD and divided wall distillation column. Dividing wall columns have gained increasing application due to their lower energy consumption and lower investment costs compared with conventional distillation column sequences. However, because of the increased design degrees of freedom, optimal process design becomes exceedingly more difficult and remains an open issue. Also, the integrated nature of the column makes the process highly interacting, and a process control system must be designed with added care [3]. Different types of divided wall distillation column are shown in Figure 1.2.

1.2 Industrial perspective of reactive distillation

The first patent for an RD process routes out in the 1920s, but little was carried out till 1980 by the Eastman Company who synthesized methyl acetate for the first time using this technique. The following reactions have shown potential for RD:

1.2.1 Esterification

In esterification reaction, alcohol and acid react to form an ester. Esters are chemical compounds having a pleasant fruity odor:

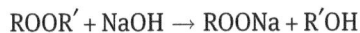

$$ROOR' + NaOH \rightarrow ROONa + R'OH$$

The main application of esters is in the synthesis of artificial flavor and essence, solvent for oil, gum, fat and resins. They are also used as plasticizers. Esterification is the oldest reaction carried out in an RD column. For example, in conventional methyl acetate production, the yield of methyl acetate is low because of low boiling azeotrope formation. This constraint is removed in RD and almost pure methyl acetate can be

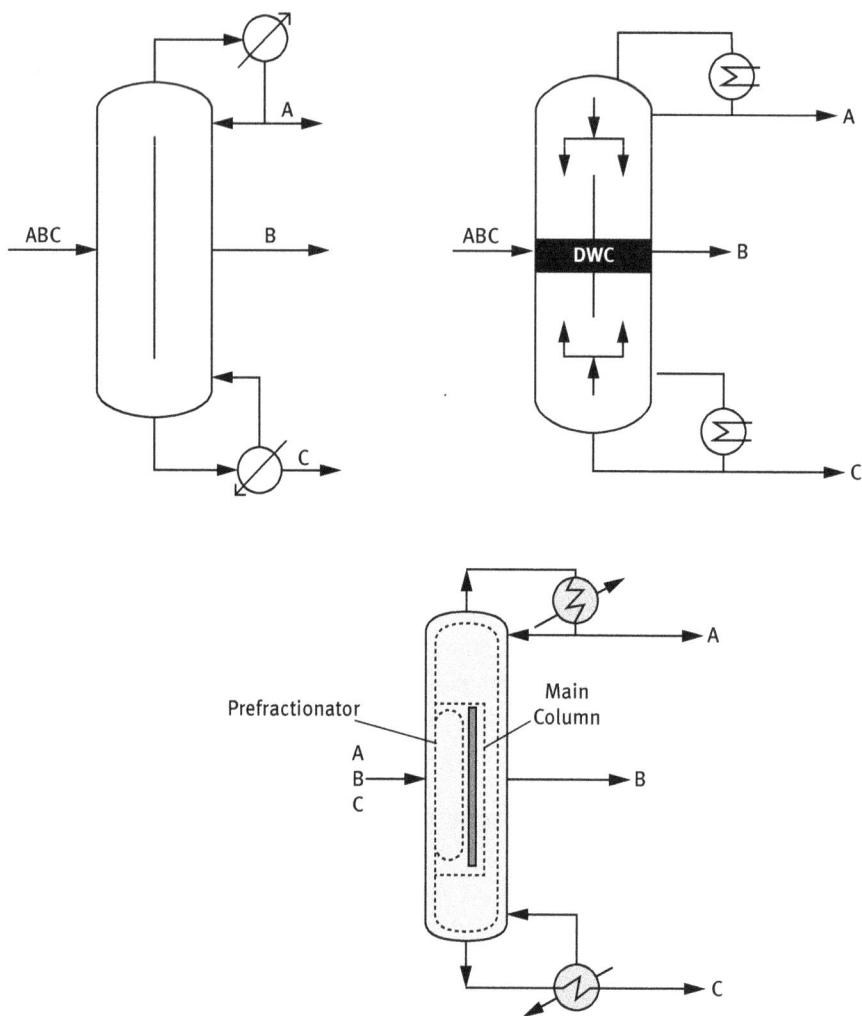

Figure 1.2: Types of reactive divided wall distillation column.

collected. Fatty acid esters are natural chemicals used, among other things in cosmetics, plastics and surfactants were also reported to be synthesized in RD [4].

In this study, we have synthesized propyl propionate by the equilibrium-limited liquid-phase esterification reaction of 1-propanol and propionic acid in a pilot-scale RD column. The standard enthalpy of reaction was experimentally determined to be −6.4 kJ/mol. This indicates that the reaction is exothermic, and the chemical equilibrium constant is slightly dependent on temperature. The reaction is not self-catalyzed and needs to be catalyzed by a strong acidic catalyst.

The reaction chemistry for propyl propionate is

$$C_3H_8O + C_3H_6O_2 \text{-----------} \rightarrow C_6H_{12}O_2 + H_2O$$

1.2.2 Transesterification

Transesterification reaction, in general, can be represented as the reaction between triglyceride and alcohol to produce alkyl esters and glycerol. The best example is a synthesis of biodiesel using transesterification. Commercially, no industrial unit has been reported on synthesis of biodiesel in RD, but the literature shows that pilot-scale synthesis is possible. This process occurs by the reaction of vegetable oil with alcohol in the presence of an alkaline or acidic catalyst as per the following reaction:

$$3\,ROOH + 3\,ROH \text{-----} \rightarrow 3\,ROOR + H_2O$$

$$\text{Fatty acid} + \text{Alcohol} \text{-----} \rightarrow \text{Alkyl ester} + \text{Water}$$

However, using methanol as the alcohol poses some problems and limitations of the process. Methanol inhibits the reaction due to the presence of water and produces free fatty acids and soap. The suitable alternative is to use ethanol instead of methanol. Ethanol is much less toxic than methanol and is readily available from the fermentation of corn. The reaction chemistry of biodiesel synthesis using ethanol and fatty acids present in vegetable oil is shown as follows:

$$3\,ROOH + 3\,ROH \text{----} \rightarrow 3\,ROOR + H_2O$$

$$\text{Fatty acid} + \text{Ethanol} \text{-----} \rightarrow \text{Alkyl ester} + \text{Water}$$

Sometimes transesterification can be a beneficial alternative to hydrolysis as it does not involve formation of water; moreover, it brings out the value added through formation of another ester. Heterogeneous catalysts are more effective from an economical point of view for biodiesel production. In this study, we have used Aberlyst-15 as a heterogeneous cation exchange catalyst for the transesterification of biodiesel using vegetable oil and ethanol.

1.2.3 Etherification

Etherification refers to the synthesis of ethers from alcohol and acid. Ethers are indispensable part of the fuel industry as like the properties of alcohol, ether also enhances the octane value of fuel when added in appropriate proportion. RD has been successfully applied to the etherification reaction to produce fuel ethers such as methyl tert-butyl ether (MTBE), tert-amyl methyl ether, ethyl tert-butyl ether (ETBE), and tert-amyl ethyl ether (TAEE). These have been the model reactions for the studies in RD in the last two decades.

These fuel oxygenates are formed by the reaction of isobutylene with alcohol to give ether and water. However, another alternative is to react tert-amyl alcohol (TAA) with the corresponding lower alcohol such as methanol or ethanol to produce ether:

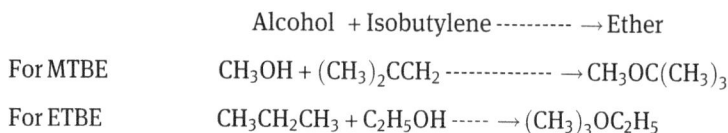

$$\text{Alcohol} + \text{Isobutylene} \cdots\cdots \rightarrow \text{Ether}$$

For MTBE $\qquad CH_3OH + (CH_3)_2CCH_2 \cdots\cdots\cdots \rightarrow CH_3OC(CH_3)_3$

For ETBE $\qquad CH_3CH_2CH_3 + C_2H_5OH \cdots\cdots \rightarrow (CH_3)_3OC_2H_5$

1.2.4 TAEE synthesis

TAEE is an excellent blending component due to its low volatility and high octane number. It does not contain aromatics and olefins, and is formed by the etherification reaction of TAA and ethanol in the presence of an acidic cation exchange catalyst, which may be either homogeneous or heterogeneous in nature.

The chemical reaction is as follows:

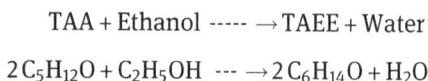

$$\text{TAA} + \text{Ethanol} \cdots\cdots \rightarrow \text{TAEE} + \text{Water}$$

$$2\,C_5H_{12}O + C_2H_5OH \cdots \rightarrow 2\,C_6H_{14}O + H_2O$$

In this study, we have synthesized TAEE as per the above reaction chemistry in a pilot-scale RD column using Aberlyst-15 as a heterogeneous catalyst [1–8].

1.3 Measurements in process industry

Measurement refers to assigning a numerical value to the physical characteristic of a system like length and diameter, and higher level properties like electrical properties and mechanical properties. On the industrial level, measurement term is scaled up to a level to measure higher order characteristic of the system such as temperature, pressure and flow rate. Any property is categorized based on the magnitude and unit which make it possible to compare it with other properties. The specification on limit of a measure is described by standard organizations like International System of Units, British thermal units, and British system of English unit. Various hardwares are attached on industrial process lines to measure these physical variables like pressure gauge and thermometer. However, hardwares are now being replaced by soft sensor techniques due to some of their demerits. For example, hardware values are intermittent in nature, hardwares need timely calibration, and they may fall trip because of the nature of material of construction, whereas soft sensors provide continuous data with more accurate results.

1.3.1 Hardware measurement

Hardware is a physical component installed in process engineering for controlling, recording, and sensing physical parameters of a machine. For example, pressure gauge is a hardware installed in process industry to give a measurement of pressure. Hardwares are expensive, time-consuming and sometimes even unreliable to sense a physical measurement. For example, to measure the quality of a product steam in terms of its composition or purity, physical sensors take long time to evaluate, leading to a too high fraction of time by keeping the next product on hold for second quality production. Thus, to have a continuous estimate of quality or such other parameters, soft sensors are now being employed.

1.3.2 Types of hardware measurements

In process industries, accurate measurements of physical quantities are essential. Common physical quantities that are used to measure hardware measurements are temperature, pressure, flow, and composition.

- **Temperature measurement:** A thermocouple is widely used for the measurement of temperature in industries. A voltage is generated when the temperature of a probe is changed because of the property that it is made of two dissimilar conductors contacting at one or more points. Thermocouple is used as a differential device rather than an absolute measurement device.
- **Pressure gauge:** A pressure gauge consists of a dial in which the scale of unit is applied, and one can read the value of pressure because of the movement of the needle inside the dial over the specified scale. Pressure gauge generally gives the value of gauge pressure which is the difference between two pressures. It can be installed easily over a vessel or a pipeline or other easy to human access areas.
- **Flow measurement:** Flow of process fluid is an indispensable part of any industry. The flow of flowing fluid is a very important design parameter which must be measured because other physical components depend directly or indirectly on this value. For example, the choice of a pump in an installed line is done based on the hydrodynamic property of flowing fluid. Variable flow can be best measured using a rotameter. The rotameter consists of a vertical tube with a metering float which is free to move within the tube. Fluid flow causes the float to rise in the tube as the upward pressure differential and buoyancy of

the fluid overcome the effect of gravity. The float rises until the annular area between the float and tube increases sufficiently to allow a state of dynamic equilibrium between the upward differential pressure and buoyancy factors, and downward gravity factors.

– **Composition measurement:** Separation and analyses of chemical components are carried out using various composition analyzers without disturbing the mixture percentage, and physical and chemical parameters. Gas chromatography is one of the most sophisticated analyzers for composition measurement. It involves vaporization of a sample which is then injected onto the head of the chromatographic column. A secondary inert or mobile gas-phase transports the sample through the column. The column itself contains a liquid stationary phase which is adsorbed onto the surface of an inert solid. There are certain detectors based on the selectivity of the sample. A nonselective detector responds to all the compounds except the carrier gas, a selective detector responds to a range of compounds with a common physical or chemical property, and a specific detector responds to a single chemical compound. A signal is generated and reported according to the Beer–Lambert law. Hence, corresponding to a wavelength, one can check the absorbance and identify the nature of compound effectively [5].

1.3.3 Importance of hardware measurement

In chemical process industry, various data are required for carrying out a reaction in an optimized manner with an objective of maximum yield or product purity. The rate of a reaction depends on various parameters like the temperature, pressure, nature of catalyst, and residence time; hence, these parameters are critical to examine. There are various types of reactions encountered in chemical engineering, which is majorly categorized as reversible and irreversible reactions. The rate of forward or backward reactions in the case of reversible reactions is a most important task to be maintained, as the equilibrium conversion is highly dependent on temperature and other parameters. Thus, there is an indispensable need of measurement of operating parameters, and data to identify and control the parameters with respect to the aim of reaction. Data collected from hardware measurement also help in developing expressions for intrinsic and the global rate of reactions, that is, chemical kinetics of the system. Properties like temperature and composition not only vary with time but may also vary with position in the reactor. Hence, hardware sensors are applied at different locations to observe the changes in these parameters and their effect on the product composition and yield. Further for a heterogeneous system consisting of solid catalyst, the specific rate constant of the reaction depends on the concentration and nature of the catalytic substance, which concludes that all these parameter measurements are very important, and a preliminary step has to be done along with continuous measurement to maintain the reaction at optimum output. These parameter measurements also reduce the risk of physical and chemical hazard in the working place. With recent advances, data measurement is also needed to be fed to online simulators and controllers to maintain the input within safe limits.

1.3.4 Difficulty in measurement

Physical sensors or hard sensors used in industries are now being replaced by soft sensors because physical sensors are expensive, time-consuming, and sometimes even unreliable to sense a physical measurement. For example, to measure the quality of a product steam in terms of its composition or purity, physical sensors take a long time to evaluate, leading to a too high fraction of time by keeping the next product on hold for second-quality production. Thus, to have a continuous estimate of the quality or such other parameters, soft sensors are now being employed. Physical sensors are also prone to have affected by the nature of chemical used in the process and hence their life is limited to be used and regular preventive maintenance or timely replacement of old hardware is required to smoothly carry out the process. Loosening of wire with time also led the hardware in nonworking condition and hence its maintenance is a very important issue.

1.3.5 Problems associated with measurement

Measurement of process variable requires the continuous operation of hardware by which system can be fully monitored. However, hardware sensors provide discrete data which need to be further studied with respect to other operating variables. This is a tedious task and requires lots of time. Secondly, measurement is always associated with certain error which cannot be rectified permanently. Various errors associated are classified as human error, system error, time lag, consistency error, and absolute error. Time lag can be defined as the time needed by the hardware to read and predict the next value and is a very critical parameter to be handled as any changes during time lag is not indicated by the hardware. Other mechanical effects like creeping and fatigue also result in mechanical degradation of hardwares which must be rectified or replaced; otherwise, it will result in increased error.

1.3.6 Suggestion/improvement in hardware sensor

Proper choice of material of construction for hardware according to the chemicals and operating condition involved is a preliminary need of designing of hardware.

Proper casing of hardware should be provided to let it remain unaffected of the mechanical wear and tear problems. Proper calibration must be done time to time to have an estimate of the error persisting in the system. This error must be included in the calculation to get a real solution of value of the measured parameter. One must study the kinetics and other physical phenomenon of mass and heat transfer associated with the process to install the sensors at accurate place where the input variable influence is maximum over the resulting output. Use of empirical relations can be applied to check the variation in the system as per the basic laws applied over the process. In view of this, one can also refer the charts and standard relations to validate the system performance [6].

1.4 Soft sensors

Soft sensors are also known by other names such as proxy sensors, or virtual sensors are used to estimate the physical sensor's measurement based on other process variables. However, soft sensor development requires skilled professional having adequate knowledge of process engineering to serve the customer. The input variables for soft sensors can be process inputs like feed flow, temperature, or some process outputs like certain product properties. By correlating these variables, a soft sensor can work on either a black, gray, or white model which involve fundamental equations, as well as some correlations between the variables based on experimental data. In a process, the inputs are converted into output through some process. The online measured process outputs are called secondary output, whereas those process outputs which cannot be measured online are called primary outputs. Thus, initially the hard sensors predict the value of secondary output which is then fed as an input along with the process input to the soft sensor. These hard-to-measure outputs are then evaluated by soft sensor in a very less time and more accurate manner. For example, in composition analyses of a product stream in a distillation column, the temperature of system at any stage is evaluated by soft sensor which is termed as secondary output. This temperature is given as feed to soft sensor which then evaluates product composition. Soft sensor works on the past experience and hence can be trained, and it can also estimate the future value of output [7–10]. The sequence of measurement using soft sensor is as shown in Figure 1.3.

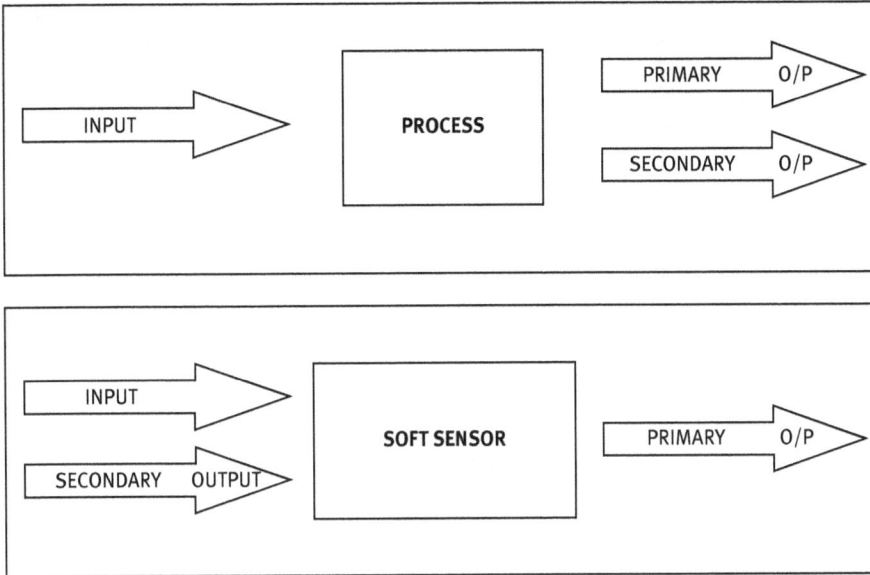

Figure 1.3: Measurement using soft sensor.

References

[1] M.P. Dudukovic, Challenges and innovations in reaction engineering, Chemical Engineering Communications, 196, 1/2, 2009, 252–266.

[2] C. Thiel, K. Sundmacher, U. Hoffmann, Synthesis of ETBE: Residue curve maps for the heterogeneously catalysed reactive distillation process, Chemical Engineering Journal, 66, 3, 1997, 181–191.

[3] D. Mohl, A. Kienle, E.D. Gilles, P. Rapmund, K. Sundmacher, U. Hoffmann, Nonlinear dynamics of reactive distillation processes for the production of fuel ethers, Computers & Chemical Engineering, 21, Supplement, 1997, S989–S994.

[4] G. Fernholz, S. Engell, L.-U. Kreul, A. Gorak, Optimal operation of a semi-batch reactive distillation column, Computers & Chemical Engineering, 24, 2–7, 2000, 1569–1575.

[5] S. Horstmann, T. Pöpken, J. Gmehling, Phase equilibria and excess properties for binary systems in reactive distillation processes: Part I. Methyl acetate synthesis, Fluid Phase Equilibria, 180, 1–2, 2001, 221–234.

[6] M.J. Olanrewaju, M.A. Al-Arfaj, Development and application of linear process model in estimation and control of reactive distillation, Computers & Chemical Engineering, 30, 1, 2005, 147–157.

[7] H.-Y. Lee, H.-P. Huang, I.-L. Chien, Design and Control of Homogeneous and Heterogeneous Reactive Distillation for Ethyl Acetate Process, Editor(s): W. Marquardt, C. Pantelides, Computer Aided Chemical Engineering, Elsevier, Vol. 21, 2006, 1045–1050.

[8] S.M. Khazraee, A.H. Jahanmiri, Composition estimation of reactive batch distillation by using adaptive neuro-fuzzy inference system, Chinese Journal of Chemical Engineering, 18, 4, 2010, 703–710.

[9]   A.K. Jana, S. Banerjee, Neuro estimator-based inferential extended generic model control of a reactive distillation column, Chemical Engineering Research & Design, 130, 2018, 284–294.

[10] S.R. Vijaya Raghavan, T.K. Radhakrishnan, K. Srinivasan, Soft sensor-based composition estimation and controller design for an ideal reactive distillation column, ISA Transactions, 50, 1, 2011, 61–70.

Chapter 2
Synthesis in reactive distillation column

Synthesis in reactive distillation column (RDC), especially in case for equilibrium-limited reaction system, is not only economical but also suitable for various complex vapor–liquid interaction systems. A wide variety of processes, including alkylation, acetylation, amination, dehydration, esterification, isomerization, hydrodesulfurization, etherification, and polymerization, are possible to perform. This reactive distillation is also suitable for multicomponent azeotropic distillation, especially heterogeneous azeotrope, because it consists of both reactive and nonreactive sections which facilitates separation of these complex azeotropes in the form of distillate.

2.1 Experimental synthesis

Experimental synthesis can be performed in laboratory-scale setup using homogeneous and heterogeneous catalyses. Glass column of various sizes can be used. The homogeneous catalyst normally used for synthesis in RDC is sulfuric acid. The number of heterogeneous catalysts can be used, including Amberlite, Amberlyst 15, and Indion [1].

Case studies
Pilot-scale-packed RDC experimental setup
Few case studies of synthesis in reactive distillation are described as below. The experiment can be performed in pilot-scale-packed RDC at ambient pressure. The temperature varies depending upon the system. The details of the pilot RDC is as below.

The internal diameter of the column is 50 mm, and the total height of the column is 3 m. The RDC consists of seven sections: among these, four sections are 50 cm long, two sections are 30 cm long, and one section is 40 cm long. Each section (rectifying section, reactive section, and stripping section) of the column is 1 m high. The nonreactive sections (rectifying and stripping) of the column are packed with noncatalytic HYFLUX (Evergreen Ltd., Mumbai). HYFLUX is high efficiency and low pressure drop wire mesh distillation packing made up of stainless steels 304 and 316. The reactive ssection is packed with KATAPAK-S (Sulzer Ltd., Switzerland) steel wire mesh packing. KATAPAK – S (S implies sandwich) is a structured catalyst support for use in gas–liquid reaction system in which catalyst granules can be embedded. KATAPAK-S is immobilized between two sheets of metal wire gauze, forming sandwiches. Each of these sheets is corrugated, resulting in a structure with flow channels of a defined angle and hydraulic diameter. Catalyst particles are filled inside the sandwich. The catalyst selected for this RDC is Amberlyst 15 (acidic ion-exchange catalyst). The heat input is through heating a mantle which acts as a reboiler. The Reboiler is mounted at the bottom of the column while condenser is fitted at the top of the column. The capacity of the reboiler is 2 L. The column is equipped with calibrated thermocouples and rotameter at various locations along the length of the column [1]. The pictorial view of RDC is shown in Figure 2.1. The column specification is listed in Table 2.1.

https://doi.org/10.1515/9783110656268-002

Figure 2.1: Pictorial view of pilot-scale reactive distillation column [1].

Table 2.1: Reactive distillation column specifications [1].

Parameter	Value
Height of column	3 m
Diameter of column	10 in
Reactive zone	From stage 3 to 6
Stripping zone	From stage 6 to 9
Rectifying zone	From stage 1 to 3
Length of reactive zone	1 m
Packing used in reactive zone	Amberlyst 15
Capacity of reboiler	2 L
Type of condenser	Vertical, total condenser
Type of reboiler	Jacketed round flask, externally heated
Maximum reboiler duty	3 kW
Number of rotameters used = 3	2 for feed flow and 1 for reflux flow control

2.1.1 Synthesis of biodiesel in a reactive distillation column

Transesterification reaction of biodiesel production can be carried out in a conventional batch reactor for homogeneous catalysis because the catalyst used is in the same phase as the reactants. However, we have carried out the same reaction using a heterogeneous catalyst in an RDC using oil and alcohol as a feed and solid Amberlyst 15 as a catalyst [2–15]. Transesterification reaction between oil and methanol can be represented as follows:

The reaction kinetic is as follows:

$$r = m_{cat}(k_1 \times x_{BD} \times x_G - k_2 \times x_{oil} \times x_{EtOH})$$

2.1.1.1 Raw materials

Mustard oil: It is the usual mustard oil available in market, which is also known as Kachi Ghani mustard oil. Physically, it is clear, free from rancidity, suspended or foreign matter, separated water, added coloring or flavoring substances or mineral oil. The specifications of the mustard oil are tabulated in Table 2.2.

Methanol: Physically, it is a colorless liquid having odor of alcohol. It should be handled with care because of highly flammable nature. Its chemical properties are listed in Table 2.3.

Amberlyst 15: It is a strong acid cation exchange resin of dark brown color. It is in the form of spherical beads. Its advantages include high catalytic activity and selectivity; resistance to fouling; long lifetime; superior resistance to thermal, mechanical, and osmotic shock; excellent stability; and low leaching, suitable for aqueous and nonaqueous media applications. Its physical properties are tabulated in Table 2.4.

2.1.1.2 Experimental procedure

The pilot-scale RDC consists of a vertical glass tube, which is divided into three zones: reactive zone, stripping zone, and rectifying zone. Each zone is provided with side taps for withdrawal of any side stream. At the bottom of the column, a round bottom reboiler is attached which is heated externally, provided by a jacket in which the heating oil is filled.

At the upper part of the column, there is a vertical condenser attached with an internal coil in which the cooling water is passed, and the product vapor gets condensed and accumulated in the condenser. A recycling tube is connected on the upper plate to provide reflux to the column. The magnetic drive pumps are used for both the feeds and reflux. A magnetic drive pump uses a balanced magnetic field to create the rotation of the fluid impeller. Unlike a traditional centrifugal pump which has a direct drive connection between impeller and motor, a magnetic drive pump eliminates the direct drive mechanism and replaces it with a magnetic field.

Outer magnetic bell housing is mounted on the end of the pump shaft. This outer bell is aligned on the outside of the rear casing. The pump impeller is connected to a smaller magnet assembly and rides on an internal shaft and bushing assembly. The smaller magnet assembly is mounted within the center of the magnetic field of the outer bell housing. Although these two magnet assemblies are separated by a fluid barrier, the magnetic fields are aligned. When the pump motor is started, the outer bell housing begins to rotate. As the outer bell rotates, the rotating magnetic field affects the inner impeller magnet. As the two magnets begin to turn together, the impeller begins turning, displacing fluid.

Oil and methanol were taken as the feed and were preheated in a feed tank to reduce their viscosity and enhance the rate of reaction and formation of products. Reboiler duty was given to provide the temperature in the column. Various temperature sensors were used to record the temperature of various segments continuously to observe the steady-state conditions and to study the temperature profile of the column.

Table 2.2: Specification of mustard oil.

Properties	Values
Quantity purchased	5 L
Acid value as % oleic acid	1.5 maximum
Iodine value by the Wij method	98–110
Turbidity temperature	23–27.5 °C
Saponification value	169–177
Unsaponifiable matter %	1–2
Refractive index at 40 °C	1.4646
Quantity used in experiment	2 L

Table 2.3: Properties of methanol [1].

Properties	Values
Quantity purchased	50 L
Chemical formula	CH_4O
Purity (%)	99%
Density (kg/m³)	792
Molecular weight (g/mol)	32

Table 2.3 (continued)

Properties	Values
Boiling point (°C)	64 °C
Quantity used in experiment	3 L

Table 2.4: Properties of Amberlyst 15 [1].

Parameters	Unit	Value
Concentration of active site	eq/kg	≥4.7
Maximum operating temperature	°C	150
Water retention capacity	%	52–57
Uniformity coefficient	Unitless	≤1.7
Fines content	Mm	Max. 1%
Coarse beads	Mm	Max. 5%
Physical form	Not applicable	Opaque beads
Surface area	m^2/g	53
Average pore diameter	Å	300
Quantity used in column	g	700

The step-by-step procedure is discussed below:

1. Cleaned the feed tank and reboiler and checked that the rotameter, condenser, and pumps are working properly.
2. Filled the feed tank with the corresponding alcohol and oil.
3. Filled the chiller and jacket of feed tank with water.
4. Filled the reboiler approximately 2.5 L with lower boiling point (more volatile) reactant, that is, methanol. Now started the heater for reboiler duty of 2 kW and set temperature of feed tanks to around 65 °C.
5. Set the reboiler temperature closer to the boiling point of component which is filled in reboiler, that is, around 70 °C to volatilize the methanol.
6. Similarly set the feed temperature lower than the boiling point of the respective component.
7. Temperature sensors reading was recorded and observed the column temperature.
8. At this point, feed of the higher boiling point component started at a flow rate of 0.02 L/min, that is, mustard oil feed at the top of reactive zone when the temperature above this section is approximately closer to the boiling point of oil.
9. Reaction proceeded and product formed is collected at the top.

10. Sample collected and analyzed for concentration measurement.
11. At this point, reflux flow to the column is started and again let the system to come under steady state so that the product starts collecting at the top. Again, a sample is taken to check the purity of the product after reflux flow [1–5].

2.1.1.3 Data logger

Data logger can be used to store data of the temperature sensors incorporated at each stage of RDC. Temperature controllers are used to control the stage temperatures as per the reboiler duty. Data logger is a data acquisition device such as plug-in boards or serial communication systems which use a computer as a real-time data recording system. However, most instrument manufacturers consider a data logger a stand-alone device that can read various types of electrical signals and store the data in internal memory for later download to a computer. The advantage of data loggers is that they can operate independently of a computer, unlike many other types of data acquisition devices. Data loggers are available in various shapes and sizes. The range includes simple economical single channel fixed function loggers to more powerful programmable devices capable of handling hundreds of inputs.

2.1.1.4 Separating funnel

It is used to collect the product from the column which separates the components of the product which may contain biodiesel as well as some glycerol or other impurities into two immiscible phases based on difference in densities.

2.1.2 Synthesis of *tert*-amyl ethyl ether (TAEE)

tert-Amyl ethyl ether (TAEE) is an excellent blending component due to its low volatility and high octane number, and they do not contain aromatics and olefins. So, the use of TAEE as environmentally friendly and economic oxygenates is increasing in recent years.

TAEE is a more favorable choice compared to TAME as one of the reactant ethanols can be derived renewably from agricultural product such as sugarcane and potatoes, while the other reactant TAA (*tert*-amyl alcohol) is a main component in fuel oil is a byproduct of fermentation process for ethanol production and can be an alternative reactant instead of iso-amylenes. Therefore, the synthesis of TAEE from TAA and ethanol can be a promising route as both reactants are derived from renewable resources:

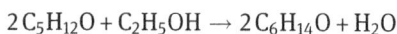

$$2\,C_5H_{12}O + C_2H_5OH \rightarrow 2\,C_6H_{14}O + H_2O$$

It can also be manufactured using iso-amylenes in place of TAA but because of easy availability of TAA in the form of by-product from other processes, iso-amylenes are seldom used [2–4].

The chemical reaction and kinetic are as follows:

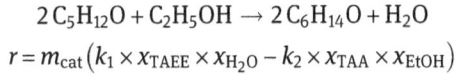

$$2\,C_5H_{12}O + C_2H_5OH \rightarrow 2\,C_6H_{14}O + H_2O$$

$$r = m_{cat}\left(k_1 \times x_{TAEE} \times x_{H_2O} - k_2 \times x_{TAA} \times x_{EtOH}\right)$$

where m_{cat} is the mass of catalyst, T is the temperature, k_1 is the forward rate constant, k_2 is the backward rate constant, and x is the concentration of components.

Chemicals used

For this, ethanol and TAA can be used to prepare an improved fuel oxygenate, that is, TAEE in the presence of heterogeneous catalyst named Amberlite 10 dry.

Ethanol: Ethanol can be derived renewably from agricultural products such as sugarcane and potatoes. In this study, the ethanol is purchased from local purchase. Physically it is a colorless liquid and its chemical properties are listed in Table 2.5.

TAA: This is the main component in fuel oil, is a by-product of fermentation process for ethanol production, and can be an alternative reactant instead of iso-amylenes. However, in this study, it is purchased from local market. It is also colorless in liquid, and its chemical properties are listed in Table 2.6.

Amberlite 10 dry: It is a dry cation exchange resin designed for the purification of biodiesel fuels produced from mineral base-catalyzed processes. It is used for removal of salts, soaps, and residual glycerol. Its physical properties are tabulated in Table 2.7.

Coolant (water): The most common coolant is water. Its high heat capacity and low cost make it a suitable heat-transfer medium. Water was used as a coolant in the condenser to condense the upcoming water vapor into a manageable liquid phase. The properties of coolant used are listed in Table 2.8.

2.1.3 Synthesis of methyl acetate

Methyl acetate, also known as acetic acid methyl ester or methyl ethanoate, is a carboxylate ester with the formula CH_3COOCH_3. It is a flammable liquid with a characteristically pleasant smell reminiscent of some glues and nail polish removers. Methyl acetate is occasionally used as a solvent, being weakly polar and lipophilic, but its close relative ethyl acetate is a more common solvent, which is being less toxic and less soluble in water. It is synthesized using acetic acid and methanol as raw materials. The reaction of methyl acetate esterification is given as follows:

$$CH_3COOH + CH_3OH \rightarrow CH_3COOC_2H_5 + H_2O$$

$$r = m_{cat}\left(k_1 \times x_{meOAC} \times x_{H_2O} - k_2 \times x_{HOAC} \times x_{MeOH}\right)$$

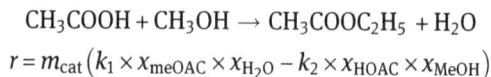

where $k_1 = k_1^0 \exp(-E_1/RT)$ and $k_2 = k_2^0 \exp(-E_2/RT)$. E_1 and E_2 are the activation energies in J/mol K. k_1^0 and k_2^0 are the standard rate constants in gmol/s, m_{cat} is the mass of heterogeneous catalyst, T is the temperature which ranges from 323 to 373 K, R is the universal gas constant (8.314 J/gmol K) [1–15].

Table 2.5: Properties of ethanol.

Properties	Values
Chemical formula	C_2H_5OH
Purity (%)	99.9
Oxygen (wt%)	34.7
Molecular weight (g/mol)	46.07
Boiling point (°C)	78
RVP (37.8 °C, kPa)	16
MON	92
RON	108
Flash point, closed cup (°C)	12.8

Table 2.6: Properties of tertiary amyl alcohol.

Properties	Values
Chemical formula	$C_5H_{12}O$
Purity (%)	99
Oxygen (wt%)	22.4
Molecular weight (g/mol)	88.15
Boiling point (°C)	101
RVP (37.8 C, kPa)	1.6
MON	100
RON	130
Flash point, closed cup (°C)	19

Table 2.7: Properties of Amberlite 10 dry.

Polymer matrix	Cross-linked polymer
Physical appearance	Brown translucent spherical beads
Fines content	<5 mm
Surface area	53
Average pore diameter	300

This reaction can be carried out either in the presence of homogeneous or heterogeneous catalyst. Heterogeneous catalyst has paramount importance in the field of industrial applications because of its several advantages such as easy separation, recovery, and larger interfacial area for the reaction to occur. In this research work, we have used Amberlyst 15 as the solid heterogeneous catalyst. Amberlyst 15 is a strongly acidic ion resin catalyst and it increases the rate of conversion of providing activation energy and large interfacial surface area for the reaction to occur.

Table 2.8: Properties of coolant.

Chemical formula	H_2O
Molecular weight (g/mol)	18
Density (4 °C, g/cm^3)	1
Boiling point (°C)	100
Latent heat of evaporation (kJ/kg)	2,270
Specific heat capacity of water (kJ/kg K)	4.187

Table 2.9: Input condition used in the synthesis of TAEE.

Parameters	Values	
Number of stages	10	
Feed flow rate	Ethyl alcohol	*tert*-Amyl alcohol
	0.03 L/min	0.03 L/min
Feed temperature	40 °C	90 °C
Feed location	Top of the reactive zone	Bottom of the reactive zone
Reboiler duty	1 kW	

Table 2.9 (continued)

Parameters	Values
Reactive zone	From stage 3 to 6
Stripping zone	From stage 6 to 9
Rectifying zone	From stage 1 to 3
Reflux ratio	3
Column packing	Katapak-S
Colum HETP	150 mm
Column diameter	10 in
Quantity of ethanol used in feed tank and reboiler	4 L
Quantity of ethanol consumed in reaction	2.5 L
Quantity of TAA in feed tank	3 L
Quantity of TAA consumed in reaction	2 L

Chemicals used

For this, methanol and acetic acid can be used to prepare methyl acetate.

Methanol: Physically, it is a colorless liquid having odor of alcohol. It should be handled with care because of highly flammable nature. Its chemical properties are listed below:

Acetic acid: It is an important chemical reagent used in the production of cellulose acetate, fiber, and fabric.

Amberlyst 15: It is a strong acid cation exchange resin of dark brown color. It is in the form of spherical beads. Its advantages include high catalytic activity and selectivity; resistance to fouling; long lifetime; superior resistance to thermal, mechanical, and osmotic shock; excellent stability; low leaching; and suitable for aqueous and nonaqueous media applications. Its physical properties are tabulated in Table 2.10.

Synthesis procedure

Acid and methanol were taken as the feed and is preheated in a feed tank to reduce their viscosity and enhance the rate of reaction and formation of a product. Reboiler duty is given to provide the temperature in the column. Various temperature sensors were used to record the temperature of various stages continuously to observe the steady-state conditions and to study the temperature profile of the column. The step-by-step procedure is discussed below:

Table 2.10: Properties of methanol.

Properties	Values
Quantity purchased	50 L
Chemical formula	CH_4O
Purity (%)	99%
Density (kg/m³)	792
Molecular weight (g/mol)	32
Boiling point (°C)	64 °C
Quantity used in experiment	3 L

Table 2.11: Properties of acetic acid.

Properties	Values
Quantity purchased	50 L
Chemical formula	CH_3COOH
Purity (%)	99.85%
Density (g/cm³)	1.05
Molecular weight (g/mol)	32
Boiling point (°C)	118 °C
Quantity used in experiment	3 L

1. Cleaned the feed tank and reboiler, and checked that rotameter, condenser, and pumps are working properly.
2. Filled the feed tank with the corresponding alcohol and acid.
3. Filled the chiller and jacket of feed tank with water.
4. Filled the reboiler approximately 2.5 L with lower boiling point (more volatile) reactant, that is, methanol. Now started the heater for reboiler duty of 2 kW and set the temperature of feed tanks to around 65 °C.
5. Set the reboiler temperature closer to the boiling point of component which is filled in reboiler, that is, around 70 °C to volatize the methanol.
6. Similarly, set the feed temperature lower than the boiling point of the respective component.
7. Sensor temperature readings are recorded and observed the column.
8. Started feeding the higher boiling point component at a flow rate of 0.02 L/ min, that is, acid feed at the top of reactive zone when the temperature above this section is approximately closer to the boiling point of acid.

9. Let the reaction to proceed and products are to be collected at the top.
10. Taken a sample and analyzed for concentration measurement.
11. Started the reflux and again let the system to come under steady state so as the product starts collecting at the top.
12. Again, taken a sample to check the purity after reflux [1–4].

2.1.4 Ethyl acetate synthesis

Ethyl acetate, also known as ethyl ethanoate, is the organic compound of ester family having a formula $CH_3COOCH_2CH_3$. It is a colorless liquid, has a characteristic sweet smell as that of methyl acetate, and is used in glues, nail polish removers, and so on. Ethyl acetate is manufactured mainly for its end use as a solvent from ethanol and acetic acid; it is manufactured on a large scale for use as a solvent. Synthesis of ethyl acetate requires ethanol and acetic acid as the reactants. These reactants react at normal temperature. However, to increase the yield reaction is usually carried out nearer to the boiling point temperature. The main product is ethyl acetate, while water is produced as a side product. Since ethanol and water form azeotropic mixture, it is required to break the azeotrope either by preventing the ethanol reactant to mix with water or by continuously removing the side product:

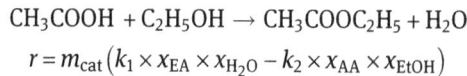

$$CH_3COOH + C_2H_5OH \rightarrow CH_3COOC_2H_5 + H_2O$$

$$r = m_{cat}\left(k_1 \times x_{EA} \times x_{H_2O} - k_2 \times x_{AA} \times x_{EtOH}\right)$$

where m_{cat} is the mass of a catalyst, T is the temperature in K), R is the universal gas constant $= 8.314$ J/gmol K, E_1 and E_2 are the activation energies in kJ/mol, $k_1{}^0$ and $k_2{}^0$ are the standard rate constants in mol/g.cat.s [5–15].

Table 2.12: Properties of Amberlyst 15.

Parameter	Value	
Concentration of active site	eq/kg	≥4.7
Maximum operating temperature	°C	150
Water retention capacity	%	52 to 57
Uniformity coefficient	Unitless	≤1.7
Fines content	Mm	Max. 1%
Coarse beads	Mm	Max 5%
Physical form	Not applicable	Opaque beads
Surface area	m^2/g	53

Table 2.12 (continued)

Parameter	Value	
Average pore diameter	Å	300
Quantity used in column	Gram	700

2.1.5 Propyl acetate

Propyl acetate is also known as propyl ethanoate. It is a colorless liquid having a characteristic odor of pears. Due to this fact, it is commonly used in fragrances and as a flavor additive. It is formed by the esterification of acetic acid and 1-propanol, often via Fischer–Speier esterification. Heterogeneous catalyzed esterification of propionic acid (ProAc) with propanol (POH) to propyl-propionate and water can be represented as follows:

$$C_3H_8O + C_3H_6O_2\text{---} \rightarrow C_6H_{12}O_2 + H_2O$$

At any time, the rate of formation of propyl propionate given for a homogeneous catalyst is used:

$$r_{propro} = K_1 C_{POH} C_{ProAc} - K_{-1} C_{propro} C_{H_2O}$$

Table 2.13: Operating condition for ethyl acetate in pilot-scale RD.

Parameters	Value
Pressure	Atmospheric
Reboiler duty	1 kW
Feed 1 (ethanol)	At the bottom of reactive zone
Feed 2 (acetic acid)	At the top of reactive zone
Feed 1 temperature	70 °C
Feed 2 temperature	85 °C
Feed 1 flow rate	0.03 L/min
Feed 2 flow rate	0.03 L/min
Reflux rate	0.25 L/min
Distillate flow rate	0.05 mL/min
Reflux ratio	5
Initial reboiler holdup	1 L

Table 2.13 (continued)

Parameters	Value
Maximum reboiler capacity	2 L
Total segment	8
Reboiler	10th stage
Condenser type	Vertical
First time to steady state	25 min
Second time to steady state	55 min
Final time to steady state	75 min
Height of column	3 m
Packing in reactive zone	Amberlyst 15 wet
Packing in rectifying + stripping zones	Katapak S

When the rate of reaction of propyl propionate is zero, it means that the equilibrium is attained, and the rate of formation is equal to the rate of disappearance:

$$k_1 C_{POH} C_{PA} = -K_{-1} C_{pp} C_{H_2O}$$
$$K_{eq} = k_1/k_2 = C_{pp} C_{H_2O} / C_{POH} C_{PA}$$

When a heterogeneous catalyst is used, the rate is affected by the amount of catalyst used [13–15].

Table 2.14: Experimental temperature with time for ethyl acetate.

Time ↓ Stage (temp) →	T1	T2	T3	T4	T5	T6	T7	T8	T9	Reboiler temp
5 min	32	34	37	38	39	38	35	35	35	65
10 min	32	34	37	37	38	35	36	34	33	65
15 min	65	35	36	37	38	36	36	34	35	64.5
20 min	69	70	80	38	38	35	36	31	35	65
25 min	71	72	82	82	81	74	75	77	77	65
30 min	71	72	82	84	80	75	76	76	79	65.5
35 min	72.5	72	82	84	81	75	76	77	77	65.5

Table 2.14 (continued)

Time ↓ Stage (temp) →	T1	T2	T3	T4	T5	T6	T7	T8	T9	Reboiler temp
40 min	72	74	80	83	80	74	75	76	80	67
45 min	72	73	81	83	79	75	78	75	79	68
50 min	73	75	80.5	82	80	76	76	76	78	68.5
55 min	74	76	81	82	79	75	76	77	80	69
60 min	74	76	81	81	79	75	75	77	80	70
65 min	75	77	80	82	77	77	75	76	82	72
70 min	75	79	81	81	78	76	76	76	81	72
75 min	74	79	80	80	76	75	76	77	81	72
80 min	75	79	80.5	81	77	75	76	76	81	72
85 min	75	78	80	81	77	76	76	76	80	72.5
90 min	75	79	80	81	77	76	76	75	80	72.5

2.1.6 Methyl *tert*-butyl ether (MTBE)

Methyl *tert*-butyl ether (MTBE) is also known as tertiary butyl methyl ether having a formula $(CH_3)_3COCH_3$, and is used as an additive for gasoline in the form of fuel oxygenate. It is volatile, flammable, and colorless liquid with little solubility in water. MTBE is a gasoline additive, which is used as an oxygenate to raise the octane number. Its use is controversial and declining because of its tendency of occurrence in groundwater which is to a certain amount carcinogenic in nature:

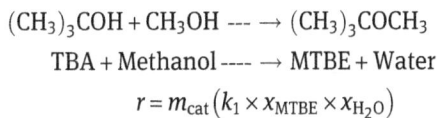

$$(CH_3)_3COH + CH_3OH \text{ --- } \rightarrow (CH_3)_3COCH_3$$
$$TBA + Methanol \text{ ---- } \rightarrow MTBE + Water$$
$$r = m_{cat}\left(k_1 \times x_{MTBE} \times x_{H_2O}\right)$$

where m_{cat} is the mass of the catalyst, T is the temperature in K, R is the universal gas constant = 8.314 J/gmol K, E_1 is the activation energy in kJ/mol, k_1^0 is the standard rate constant in L/mol min [8–15].

2.1.7 ETBE synthesis in pilot RD

Ethyl *tert*-butyl ether (ETBE) is a fuel oxygenate of ether family, commonly used as an oxygenate gasoline additive which adds an extra oxygen to make the combustion complete and is also used as a raw material in the production of gasoline from crude oil [6–15]. ETBE increases the octane number and does not cause any vapor-locking problem in engine like that of ethanol:

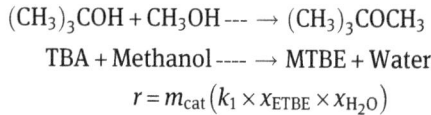

$$(CH_3)_3COH + CH_3OH \text{ --- } \rightarrow (CH_3)_3COCH_3$$
$$TBA + Methanol \text{ ---- } \rightarrow MTBE + Water$$
$$r = m_{cat}\left(k_1 \times x_{ETBE} \times x_{H_2O}\right)$$

Table 2.15: Operating condition for propyl acetate.

Parameters	Value
Pressure	Atmospheric
Reboiler duty	2 kW
Feed 1 (propanol)	At the bottom of reactive zone
Feed 2 (acetic acid)	At the top of reactive zone
Feed 1 temperature	85 °C
Feed 2 temperature	90 °C
Feed 1 flow rate	0.03 L/min
Feed 2 flow rate	0.03 L/min
Reflux rate	0.25 L/min
Distillate flow rate	0.05 mL/min
Reflux ratio	5
Initial reboiler holdup	1 L
Maximum reboiler capacity	2 L
Total segment	8
Rectifying section	1–3
Reactive zone	3–6
Stripping section	6–9
Reboiler	10th segment
Condenser type	Vertical
Height of the column	3 m

Table 2.15 (continued)

Parameters	Value
Packing in the reactive zone	Amberlyst 15 wet
Packing in rectifying + stripping zones	Katapak S

Table 2.16: Operating condition for MTBE.

Parameters	Value
Pressure	Atmospheric
Reboiler duty	2 kW
Feed 1 (methanol)	At the bottom of reactive zone
Feed 2 (Tba)	At the top of reactive zone
Feed 1 temperature	60 °C
Feed 2 temperature	75 °C
Feed 1 flow rate	0.02 L/min
Feed 2 flow rate	0.02 L/min
Reflux rate	0.25 L/min
Distillate flow rate	0.05 mL/min
Reflux ratio	5
Initial reboiler holdup	1 L
Maximum reboiler capacity	2 L
Total segment	8
Reboiler	10th stage
Condenser type	Vertical
First time to steady state	20 min
Second time to steady state	50 min
Final time to steady state	75 min
Height of column	3 m
Packing in reactive zone	Amberlyst 15 wet
Packing in rectifying + stripping zones	Katapak S

Table 2.17: Experimental temperature data for MTBE.

Time ↓ Stage (temp) →	T1	T2	T3	T4	T5	T6	T7	T8	T9	Reboiler temp
5 min	32	34	37	38	39	38	35	35	35	65
10 min	32	34	37	37	38	35	36	34	33	65
15 min	65	35	36	37	38	36	36	34	35	64.5
20 min	55	55	54	52	55	34	33	34	35	65
25 min	55	54	55	53	55	53	31	33	35	65
30 min	56	57	55	56	55	53	31	32	33	65.5
35 min	55	56	54	56	58	54	48	45	40	65.5
40 min	60	58	53	55	58	57	51	52	51	67
45 min	65	61	63	66	69	66	65	59	57	68
50 min	54	54	55	59	60	54	56	56	58	68.5
55 min	55	56	51	52	59	55	56	57	50	69
60 min	54	56	51	51	59	55	55	57	50	67
65 min	55	57	50	52	57	57	55	56	52	65
70 min	55	59	51	51	58	56	56	56	51	65
75 min	54	59	50	50	56	55	56	57	51	65.8
80 min	55	59	50	51	57	55	56	56	51	72
85 min	55	58	50	51	57	56	56	56	50	71
90 min	55	59	50	51	57	56	56	55	50	71

2.2 Dual RD integration technique

In many chemical reaction processes like esterification, etherification, and fatty acid reactions, there are some probability of obtaining a product or by-product or a side stream which can be further treated with some input into a secondary column to execute two or more reactions at a time that eventually results into process intensification as it saves the space required, time consumed, and cost of operation [1].

Table 2.18: Experimental temperature data for ETBE.

Time (min)	T1C	T2C	T3C	T4C	T C	T6C	T7C	T8C	T9C	Reboiler
10	32	36	32	37	37	35	34	30	31	61
20	62	63	60	38	38	35	35	31	32	63
30	61	64	60	66	67	62	56	55	54	63
40	64	64	62	66	77	62	56	49	55	65
50	65	68	64	70	75	66	57	53	54	65
60	65	68	63	67	73	66	58	52	53	66
70	66	66	63	67	76	65	59	57	53	65
80	67	66	64	68	70	63	54	53	52	67
90	65	69	63	65	67	63	55	52	53	67
100	65	69	63	65	67	63	55	52	52	67

Table 2.19: Operating condition for ETBE.

Parameters	Value
Pressure	Atmospheric
Reboiler duty	2 kW
Feed 1 temperature	60 °C
Feed 2 temperature	75 °C
Feed 1 flow rate	0.02 L/min
Feed 2 flow rate	0.02 L/min
Reflux rate	0.25 L/min
Distillate flow rate	0.05 mL/min
Reflux ratio	5
Initial reboiler holdup	1 L
Maximum reboiler capacity	2 L
Total segment	8
Rectifying section	1–3
Reactive zone	3–6
Stripping section	6–9
Reboiler	10th stage

Table 2.19 (continued)

Parameters	Value
Condenser type	Vertical
First time to steady state	30 min
Final time to steady state	70 min
Height of column	3 m
Packing in reactive zone	Amberlyst 15 wet
Packing in rectifying + stripping zones	Katapak S

Thus, dual reactive distillation is a part of such technique which provides the following additional benefits:

1. Heat integrations add benefit to a system by decreasing the waste energy, consuming less input energy, and thermally coupling the process to eliminate a reboiler, a condenser, or both.

2. Petluyk column which is a representative of a dual-reactor system is a well-known configuration that overcomes the thermodynamic and chemical equilibrium limitations which maximize the conversion.

3. In view of this, we have proposed the membrane-assisted reactive divided wall (MRDW) distillation for the blending of biodiesel with ether fuel oxygenate. The proposed technique is compared with the dual RD process in terms of cost and it is found that membrane-assisted reactive divided wall configuration leads to 12% reduction in cost and process heat integration with maximum conversion by eliminating extra reboiler and condenser.

4. The production of biodiesel by transesterification in existing conventional processes requires excess alcohol, typically 100%, over its stoichiometric requirement in order to drive the chemical reaction to complete. This excess alcohol must be recovered and purified for reusing by rectification and distillation, which involves additional capital and operating costs. Therefore, combination of reactor and distillation column in only one unit called RDC may lead to an enormous capital investment cost reduction. Thus, reactive distillation as a part of process intensification is economical for equilibrium-limited reversible reactions such as biodiesel synthesis as well as etherification of TAEE. Hence, we have shown the synthesis of both biodiesel and TAEE in separate RDCs which can be then mixed for blending [1–12]. The schematic diagram is shown in Figure 2.2.

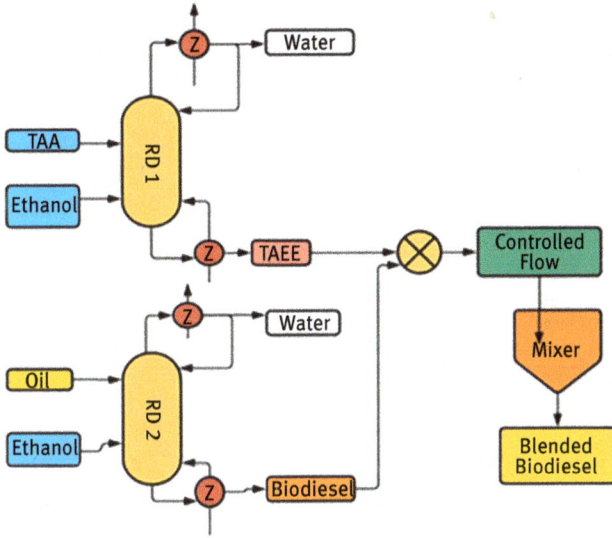

Figure 2.2: Schematic diagram of dual RD integration technique.

Table 2.20: Input condition and kinetic data for methyl acetate reactive distillation [6–12].

Parameters	Value
Pressure	Atmospheric
Reboiler duty	1 kW
Feed 1 (methanol)	At the bottom of reactive zone
Feed 2 (acetic acid)	At the top of reactive zone
Feed 1 temperature	50 °C
Feed 2 temperature	75 °C
Feed flow rate	0.02 L/min
Reflux rate	0.15 L/min
Distillate flow rate	0.05 mL/min
Reflux ratio	3
Maximum reboiler capacity	2 L
Total stage	10
Rectifying section	1–3
Reactive zone	3–6
Stripping section	6–9

Table 2.20 (continued)

Parameters	Value	
Reboiler	10th stage	
Condenser type	Vertical	
Height of column	3 m	
Packing in reactive zone	Amberlyst 15 wet	
Packing in rectifying + stripping zones	Katapak S	
Activation energy (J/mol K)	Forward 53,340	Backward 30,363
Specific reaction rate (gmol/s)	13,325	17.67
Heat of reaction (kJ/mol)	22.9	
Heat of vaporization (cal/g)	96	

2.3 Membrane-assisted reactive divided wall (MRDW) distillation process

We have proposed a novel process integration concept in which we have used a divided wall column, which consists of a wall to the center of the column composed of component-selective membranes at the two sides. This saves the cost tremendously. Membranes have numerous numbers of useful properties such as resistance to mechanical, chemical, and thermal stress, high available surface area per unit volume, high selectivity, and the ability to control the components in contact. Inorganic membranes are preferred because of their ability to withstand harsh conditions such as higher temperatures and high acidic and basic environments. Commonly used membranes include metallic, ceramic, zeolite, or carbon made in a reactive divided wall column. Literature shows that for biodiesel separation, carbon membrane is the most widely used, while zeolite membrane is best suited for TAEE separation. Further, the column consists of 10 sections and is filled with Amberlyst 15 in the middle, that is, from section 3 to 6. Ethanol is the common reactant for biodiesel and TAEE synthesis. It is filled in the reboiler at the base to attain the temperature throughout the column. Oil is supplied from one side as a reactant for biodiesel synthesis on the top of the reactive zone, while the TAA used for TAEE synthesis is supplied from the other side of the column. Another major advantage to this technique lies in the cost saving, as ethanol is not to be supplied continuously to the column, it is filled in the reboiler at the base and is recycled continuously. The upcoming ethanol vapors are directed to both the sections in the column separated by the wall and react with downcoming oil in one section while with TAA at the other. The chemical reaction leads to the production of biodiesel and TAEE, which will be redirected to a collection vessel through a flow control valve. The chapter envisaged to calculate the TAEE and biodiesel flow rate. The two were to be mixed in a proportion to achieve a target, where the final blend remains B20 (%).

The calculated % biodiesel blend as obtained from MRDW is 19.7%, which is within the legal limit.

The formula used for calculating biodiesel blend (%) is given as follows:

$$\text{Legal limit of biodiesel blend} = \left(\frac{\text{Molecular weight of the TAEE}}{\text{Molecular weight of the oxygen}} \right) \times \text{oxygen \% in blend}$$

Further, biodiesel and TAEE produced were sent to the mixer to analyze the effect of blending of fuel with fuel oxygenate. Inline blending process was adopted, where we can easily maintain the required blend grade by adjusting the biodiesel and TAEE flow rates [1–4].

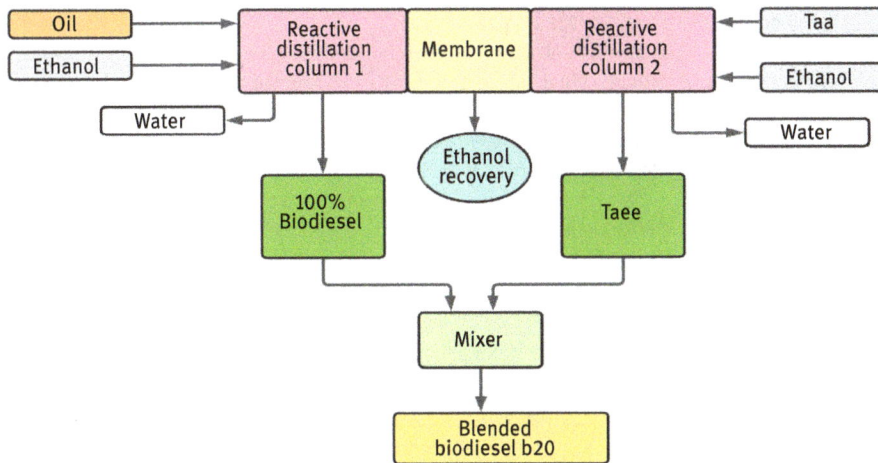

Membrane assisted dual reactive distillation column

Figure 2.3: Flow diagram of MRDW [1].

References

[1] V. Sakhre, S. Jain, V.S. Sapkal, D.P. Agarwal, Novel process integration for biodiesel blend in membrane reactive divided wall (MRDW) column, Polish Journal of Chemical Technology, 18, 1, 105–112.

[2] B. Mauro, G. Giuseppe, Nb_2O_5-catalyzed kinetics of fatty acids esterification for reactive distillation process simulation, Chemical Engineering Research & Design, Transactions of the Institution of Chemical Engineers Part A, 100, 2015, 292–301.

[3]   K.R. Dewi, A.I. Novrizal, H. Hadiyanto, B. Arief, Application of Tin (II) chloride catalyst for high FFA jatropha oil esterification in continuous reactive distillation column, Bulletin of Chemical Reaction Engineering & Catalysis, 11, 1, 2016, 66–74.

[4]   C.A. González-Rugerio, R. Fuhrmeister, D. Sudhoff, J. Pilarczyk, A. Góraka, Optimal design of catalytic distillation columns: A case study on synthesis of TAEE, Chemical Engineering Research & Design: Transactions of the Institution of Chemical Engineers Part A, 92, 3, 2014, 391–404.

[5]   M. Mallaiah, K.A. Kishore, G.V. Reddy, Catalytic reactive distillation for the esterification process: experimental and simulation, Chemical & Biochemical Engineering Quarterly, 31, 3, 2017, 293–302.

[6]   C. Chavdar, S. Evgeni, Reactive distillation for ethyl acetate production, Journal of Chemical Technology & Metallurgy, 52, 3, 2017, 463–474.

[7]   A. Amnart, Gani, R. Górak, Andrzej, A. Suttichai, Methodology for design and analysis of reactive distillation involving multielement systems, Chemical Engineering Research & Design: Transactions of the Institution of Chemical Engineers Part A, 89, 8, 2011, 1295–1307.

[8]   H. Bisowarno Budi, O. Tadé Moses, The comparison of disturbance rejection properties of one-point control schemes for ETBE reactive distillation, Chemical Engineering Communications, 189, 1, 2002, 85–111.

[9]   U. Muhammad, Y. Alhamed, A. Alzahrani, A. Saleemi, Residue curve map determination and synthesis of ethyl tert-butyl ether via reactive distillation, Arabian Journal for Science & Engineering (Springer Science & Business Media B.V.), 39, 4, 2014, 2475–2481.

[10]   A. Va Veronika, Simulation of industrial MTBE production in reactive distillation column using ASPEN HYSYS, Petroleum & Coal, 60, 1, 2018, 14–23.

[11]   E. Mityanina Olga, A. Samborskaya Marina, N. Sabiev Ayan, Mathematical modeling and steady states multiplicity analysis of methyl tert-butyl ether reactive distillation synthesis, Petroleum & Coal, 58, 4, 2016, 490–498.

[12]   H. Salvador, S-Hernández, J.G. J-Trujillo Lorena, J.E. E-Pacheco, R. Maya-Yescas, Design study of the control of a reactive thermally coupled distillation sequence for the esterification of fatty organic acids, Chemical Engineering Communications, 198, 1, 2011, 1–18.

[13]   T. Issariyakul, A.K. Dalai, Comparative kinetics of transesterification for biodiesel production from palm oil and mustard oil, The Canadian Journal of Chemical Engineering, 90, 2012, 342–350.

[14]   B. Nezahat, Kinetic studies for the production of tertiary ethers used as gasoline additives, Doctoral Thesis, The Middle East Technical University, 2004.

[15]   R.S. Huss, F. Chen, M.F. Malone, M.F. Doherty, Reactive distillation for methyl acetate production, Computers & Chemical Engineering, 27, 2003, 1855-/1866.

Chapter 3
Modeling of reactive distillation column

3.1 Modeling of reactive distillation column

In this chapter, a semirigorous mathematical model of reactive distillation process is discussed and considered for the analysis and control purpose. A semirigorous model assumes rapid energy dynamics, and the enthalpy balance equations are reduced to an algebraic equation. This means that the thermal equilibrium is achieved faster than the vapor–liquid equilibrium. The schematic diagram of the packed reactive distillation column is shown in Figure 3.1.

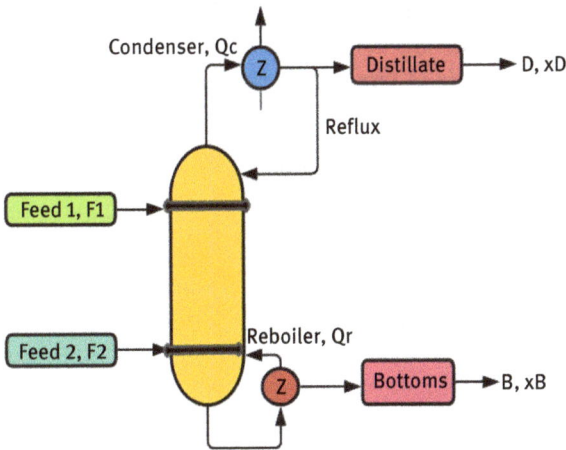

Figure 3.1: Schematic diagram of reactive distillation column.

The following modeling assumptions are made:
    Assumptions:
1) Constant relative volatility
2) Constant liquid holdup in the reactive zone, reboiler, and condenser
3) Assuming reactive zone to be a single stage
4) Negligible vapor holdup
5) Reaction to occur in the reactive zone and in the reboiler.

3.1.1 Material and energy balance around condenser

Condenser is located at the top of column in which the vapors of more volatile component are condensed to get the product in liquid form. The condensed liquid is split

https://doi.org/10.1515/9783110656268-003

into two parts at the outlet of condenser. The liquefied part returned to the column is called reflux while the product is collected as the second stream. The schematic diagram of the top section of the reactive distillation column is shown in Figure 3.2.

The overall balance around the condenser can be given as follows:

$$V_N = R + D \qquad (3.1)$$

where V_N is the vapor rate coming from top of the column, R is the reflux rate, and D is the distillate flow rate.

Similarly, component balance is made around the condenser for more volatile components which is given as follows:

$$V_N y_N = (R + D) x_D \qquad (3.2)$$

where V_N is the vapor rate coming from top of the column, y_N is the vapor fraction of more volatile components, R is the reflux rate, D is the distillate flow rate, x_D is the liquid fraction of more volatile components in distillate.

Condenser removes the heat of vapors in terms of latent heat of condensation for which energy balance is applied as follows:

$$d(M_D h_D) = - Q_C + V_N H_N - (R + D) h_D \qquad (3.3)$$

where Q_c is the cooling duty of the condenser supplied through chillers and M_D is liquid holdup in condenser. H_N is the enthalpy of vapor $= C_P T + \lambda$. h_D is the enthalpy of liquid distillate $= C_P T$, C_P is the specific heat of a distillate product, T is the temperature of distillate, and λ is the latent heat of cooling [1–9].

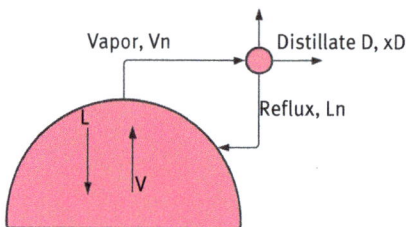

Figure 3.2: Top section of reactive distillation column.

3.1.2 Balance around reactive zone

Reactive zone in the reactive distillation column is the zone in which solid catalyst is filled and reaction occurs in this zone in the presence of a solid catalyst. Consider the feed is supplied at the nth stage in reactive zone at which the upcoming vapor and downcoming liquid interact to exchange more volatile components in the vapor. The stage numbering is done from top to bottom, that is, top plate is numbered as 1 and last is 10th. The schematic diagram of the reactive zone is shown in Figure 3.3. Balance can be applied as follows:

$$V_{n+1} + L_{n-1} + F_{n1} + F_{n2} = V_n + L_n \tag{3.4}$$

Here at the nth stage, the vapors coming upward from lower stage and liquid coming down from the above stage will work as input along with the feed, while the vapor and liquid leaving from the nth plate will work as the outgoing stream.

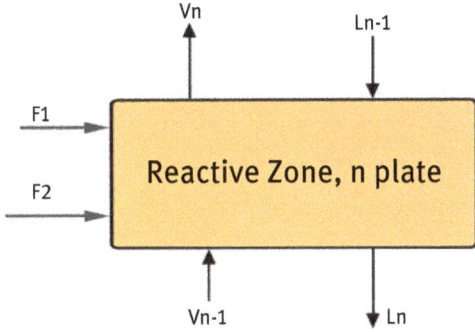

Figure 3.3: Middle section (reactive zone) of the column.

3.1.3 Component balance around reactive zone

Balance for more volatile components is applied for the nth stage. This balance equation included the reaction rate kinetic by assuming reaction to be of first order. Mathematically, it can be written as follows:

$$\frac{d}{dt}(VC_A) = V_{n+1}Y_{n+1} + L_{n-1}.x_{n-1} + F_{n1}x_{F1} + F_2x_{F2} - V_nY_n - L_n.x_n - VkC_A^1 \tag{3.5}$$

where V_{n+1} is the vapor flow rate coming as in from the lower section on the nth segment, y_{n+1} is the vapor fraction of vapor coming from the lower section on the nth segment, L_{n-1} is the liquid flow rate coming on the nth section from the upper segment, x_{n-1} is the fraction of more volatile components in liquid coming from the upper segment, F_{n1} is the feed flow rate of less volatile reactants, F_{n2} is the feed flow rate of more volatile reactants, V_n is the vapor flow rate leaving the nth segment, L_n is the liquid leaving section n, V is the volume in the reactive zone in which reaction occurs, k is the reaction rate constant, and C_A is the concentration of reactant A [1–9].

3.1.4 Energy balance around reaction zone

The reaction occurring in reactive zone may be either exothermic or endothermic in nature. Thus, the energy balance equation must accompany the energy added to the zone from the reaction or energy consumed in the reaction. The overall energy balance in reactive zone can be given as follows:

$$\frac{d}{dt}(Vh) = V_{n+1}H_{n+1} + L_{n-1}.h_{n-1} + F_{n1}h_{F1} + F_{n2}h_{F2} - V_nH_n - L_n.h_n - \lambda VkC_A^1 \qquad (3.6)$$

where V_{n+1} is the vapor flow rate coming as in from the lower section on the nth segment, H_{n+1} is the enthalpy of vapor coming from the lower section on the nth segment, L_{n-1} is the liquid flow rate coming on the nth section from the upper segment, h_{n-1} is the enthalpy of more volatile components in liquid coming from the upper segment, F_{n1} is the feed flow rate of less volatile reactants, F_{n2} is the feed flow rate of more volatile reactants, V_n is the vapor flow rate leaving the nth segment, H_n is the enthalpy of vapor leaving from the nth segment, L_n is the liquid leaving section n, h_n is the enthalpy of liquid leaving segment n, V is the volume in the reactive zone in which reaction occurs, k is the reaction rate constant, C_A is concentration of reactant A, and λ is the latent heat of vaporization.

3.1.5 Balance around reboiler

In reboiler, the initial amount of more volatile components is filled, and a heat duty is given to vaporize this liquid to provide temperature to the column. The downcoming liquid is also further vaporized from reboiler duty to convert more of more volatile liquids to be collected as the top product in the distillate. The schematic diagram of the bottom section is shown in Figure 3.4. The overall balance can be written as follows:

$$L_n = V_B + B \qquad (3.7)$$

where L_n is the liquid coming from the bottom of column into reboiler, V_B is the vapor coming from reboiler into the column, and B is the bottom product.

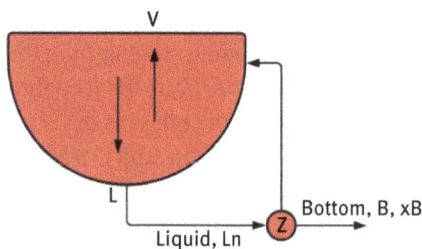

Figure 3.4: Bottom section of the reactive distillation column.

3.1.6 Component balance around reboiler

Component balance for reboiler is very simple as it involves no reaction parameter because it is assumed that the reaction occurs only in the reactive zone. Mathematically, it can be written as follows:

$$L_n x_n = V_B y_B + B x_B \tag{3.8}$$

where L_n is the liquid coming from the bottom of column into reboiler, x_n is the fraction of more volatile components in the liquid at the last segment, V_B is the vapor coming from reboiler into the column, y_B is the vapor fraction of more volatile components in vapor coming from reboiler, B is the bottom product, and x_B is the vapor fraction of more volatile components in the bottom product.

3.1.7 Energy balance around reboiler

Energy balance is very important to be applied for reboiler because the temperature in the column is mainly because of the heat duty provided from reboiler. It can be written as follows:

$$L_n h_n = V_B H_B - B h_B = \frac{d}{dt}(M_B h_B) + \theta \tag{3.9}$$

where L_n is the liquid coming from the bottom of column into reboiler, h_n is the enthalpy of liquid leaving the last plate of reboiler, V_B is the vapor coming from reboiler into the column, H_B is the enthalpy of vapor stream coming from reboiler, B is the bottom product, h_B is the enthalpy associated with the bottom product, M_B is the bottom product holdup, and θ is the latent heat of vaporization of liquid held in reboiler [1–9].

3.2 Nonequilibrium modeling

Nonequilibrium modeling is concerned with transport process and rate of reaction which change components quantity over time. The overall balance for any segment i in the column can be represented as shown in Figure 3.5.

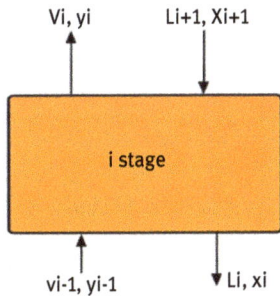

Figure 3.5: Flow quantities over stage *i*.

Applying material balance is like the one applied in equilibrium modeling. The only difference is that this equation will also consider the flux of mass transfer of more volatile and less volatile components as follows:

$$\frac{d}{dt}(M_R) = V_{i-1}y_{i-1} + L_{i+1}x_{i+1} - L_ix_i - V_iy_i - N_{Ai_L}A + N_{Ai_V}A \tag{3.10}$$

where N_A is the flux of a diffusing material, which is here considered as the more volatile components, and M_R is the reactor holdup.

The flux can also be represented using Fick's law, which states that the flux is directly proportional to the concentration gradient and mathematically it is written as follows:

$$N_A = -D_A\frac{dC_A}{dz} \tag{3.11}$$

where C_A is the concentration of component A and z is the distance. D_A is the diffusivity which is proportionality constant in Fick's law and is a function or product purity, temperature, and pressure. Thus, it can be written as $D_{ij} = f(x_i, T, P)$. Also, the flux balance over a control volume can be applied as follows:

$$(N_iA)_n - (N_iA)_{n+1} = \sum V_{i,m}R_mE_kV_n \tag{3.12}$$

where E_k is the total catalyst volume available in the contacting cell, V_n is the volume fraction of the slice of catalyst, $V_{i,m}$ is the stoichiometric coefficient of component i, N_i is the molar flux, A is the area through which diffusion occurs, and n is the plate.

Moreover, reactive distillation dealing with two reactants (A and B) producing two products (C and D) can be chemically generalized as follows:

$$A + B \leftrightarrow C + D$$

Some assumptions are made to reduce the number of design optimization:
1. The process is ideally in vapor–liquid equilibrium.
2. Saturated liquid feed and reflux flow rate.
3. The energy equations are assumed to be constant molar overflow except in the reactive zone.
4. The products have constant relative volatilities.
5. Heat of reaction and vaporization and saturated liquid feed and reflux are fixed.

The net reaction rate for component j on segment n in the reactive zone is given by

$$R_{n,j} = v_jM_n(k_{Fn}x_{n,A}x_{n,B} - k_{Bn}x_{n,C}x_{n,D}) \tag{3.13}$$

where $R_{n,j}$ is the net reaction rate, v_j is the reaction volume, M_n is the holdup in the reactive zone, k_{Fn} is the forward rate constant, and k_{Bn} is the backward rate constant.

The rate constant k can be related to activation energy as per the Arrhenius law relation:

$$\ln(k) = \left[\left(-\frac{E}{RT} \right) + \ln(A) \right] \qquad (3.14)$$

This has the same form as an equation for a straight line given as follows:

So, when a reaction has a rate constant that obeys the Arrhenius equation, a plot of $\ln(k)$ versus T^{-1} gives a straight line, whose gradient and intercept can be used to determine E_a and A. Here, E_a is the activation energy, A is the Arrhenius constant, R is the gas constant, T is the temperature, and k is the rate constant of the reaction.

The heat of reaction vaporizes some liquid on each stage within the reactive section. Therefore, the vapor rate increases up through the reactive trays and the liquid rate decreases down through the reactive trays. The vapor and liquid flow rate can also be related to the energy as follows:

$$V_n = V_{n+1} - \frac{\lambda}{\Delta H_v} R_{n,C} \qquad (3.15)$$

$$L_n = L_{n+1} + \frac{\lambda}{\Delta H_v} R_{n,C} \qquad (3.16)$$

where V_n is the vapor leaving from segment n, V_{n+1} is the vapor coming from lower section on the nth stage, λ is the latent heat of vaporization, ΔH_v is the change in enthalpy, and $R_{n,j}$ is the rate of reaction [1–9].

3.3 Dynamic balances for the column

The model equations developed are employed to simulate the dynamics of the reactive distillation column. These model equations have the main features of the reactive distillation column and represent essential dynamics of the system. The input–output flow quantities are shown in Figure 3.6. As reactive distillation exhibits nonequilibrium conditions, the following assumptions are considered:

1. Three zones in reactive distillation column: reactive, rectifying, and stripping sections
2. Constant relative volatility
3. Variable liquid and constant vapor holdup throughout the column
4. Complete mixing of vapor and liquid
5. Chemical reaction occurs in liquid phase in reactive zone and in reboiler. No reaction occurs in vapor phase.
6. Negligible pressure drops

Total condensation with no subcooling.

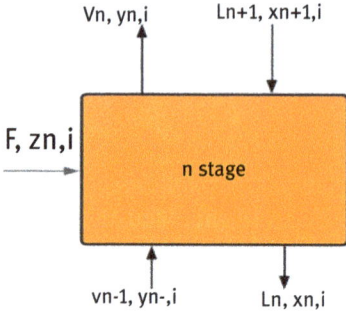

Figure 3.6: Input–output flow quantities of the *n*th stage.

3.3.1 Reflux drum

This is the topmost section of reactive distillation in which we have assumed that no reaction is taking place. The dynamic balance can be applied as follows:

$$\frac{d(M_D x_{D,j})}{dt} = V_{NT} y_{NT,j} - D(1 + RR) x_{D,j} \tag{3.17}$$

where M_D is the holdup of the reflux drum, $x_{D,j}$ is the fraction of the component in the distillate, V_{NT} is the vapor flow rate from the top stage, $y_{NT,j}$ is the fraction of the component in vapor phase, D is the amount of the distillate formed, and RR is the reflux rate.

3.3.2 Rectifying and stripping zones

These zones are located above and below the reactive zone, respectively. No reaction is assumed to be carried out in these zones. The balance can be applied as follows:

$$\frac{d(M_n x_{n,j})}{dt} = L_{n+1} x_{n+1,j} + V_{n-1} y_{n-1,j} - L_n x_{n,j} - V_n y_{n,j} \tag{3.18}$$

where M_n is the liquid holdup, L_{n+1} and V_{n-1} are rates of liquid and vapor flows, x_{n+1} and y_{n-1} are liquid and vapor fractions, respectively.

3.3.3 Reactive zone

In this zone, an extra term of reaction will be added, and hence the model equation can be written as follows:

$$\frac{d(M_n x_{n,j})}{dt} = L_{n+1} x_{n+1,j} + V_{n-1} y_{n-1,j} - L_n x_{n,j} - V_n y_{n,j} + R_{n,j} \tag{3.19}$$

where $R_{n,j}$ is the rate of reaction at the nth stage. Other notations have the same meaning as mentioned earlier.

3.3.4 Feed trays

Feed trays are those to which the feed can be supplied. In case of reactive distillation, feed is usually supplied at and bottom of the reactive zone, respectively. The component balance around the feed tray can be written as follows:

$$\frac{d\left(M_n x_{n,j}\right)}{dt} = L_{n+1} x_{n+1,j} + V_{n-1} y_{n-1,j} - L_n x_{n,j} - V_n y_{n,j} + R_{n,j} + F_n z_{n,j} \tag{3.20}$$

where F_n is the amount of feed rate at the nth stage and $z_{n,j}$ is the fraction of the jth component on the nth stage present in the feed.

3.3.5 Column base

Column base is a balance around the reboiler. This can be written as follows:

$$\frac{d\left(M_B x_{B,j}\right)}{dt} = L_1 x_{1,j} - B x_{B,j} - V_B y_{B,j} \tag{3.21}$$

where B is the amount of bottom flow rate, $x_{B,j}$ is the liquid composition of jth component in the bottom, V_B is the volume of the reboiler, and $y_{B,j}$ is the vapor composition of jth component in the reboiler.

3.3.6 Energy balance

If change in enthalpy for rectifying and stripping sections is constant, then it varies in the reactive zone because of exothermic heat of reaction involved for esterification reaction:

$$\frac{d}{dt}\left(V_R h_R\right) = V_{n+1} H_{n+1} + L_{n-1} h_{n-1} + F h_F - V_n H_n - L_n h_n - \lambda V_R R_{n,j} \tag{3.22}$$

where V_R is the volume of the reactive zone, h_R is the enthalpy of liquid in the reactive zone, V_{n+1} and L_{n-1} are vapor and liquid flow rates in and out of the nth stage, H_{n+1} and h_{n-1} are vapor and liquid enthalpies in and out of the nth stage, F is the feed flow rate, h_F is the enthalpy of the feed, H_n and h_n are vapor and liquid enthalpies at the nth stage, λ is the heat of reaction, and other notations have the same meaning are mentioned earlier.

3.3.7 Condenser heat duty

$$Q_C = V_n H_n - (R + D) h_D \qquad (3.23)$$

where Q_C is the condenser heat duty, h_D is the enthalpy of the liquid distillate, and other notations have usual meanings as already mentioned.

3.3.8 Reboiler heat duty

$$Q_B = B H_{B,j} + V_B h_{B,j} - L_n h_{n,j} \qquad (3.24)$$

where Q_B is the reboiler heat duty, $H_{B,j}$ and $h_{B,j}$ are the vapor and liquid enthalpies of the jth component in the reboiler, and other notations have their usual meanings as already mentioned.

3.3.9 Column pressure

Pressure along a column can be related to mole fraction and partial pressure of component using Roault's law as follows:

$$P = \sum_{j=1}^{NC} x_{n,j} P_{j(Tn)}^{S} \qquad (3.25)$$

And the fraction of components can be calculated as follows:

$$y_{n,j} = \frac{P_j^S}{P} x_{n,j} \qquad (3.26)$$

However, vapor pressure can be given using Antoine's equation as follows:

$$T = B/A - \log P - C \qquad (3.27)$$

where A, B, and C are Antoine's constants at temperature [5–11].

3.3.10 Temperature measurement

The temperature measurement as a secondary variable as an input for neural network estimator can be written for three stages selected as follows: the condenser and the reflux drum, the reactive stage, and bottom reboiler temperature. For condenser, the change in temperature with respect to time can be written as follows:

$$\frac{dT_n}{dt} = DRx_D \left[\frac{(T_{n+1} - T_n)}{M_c} \right] - B(1 - x_D) + \left(\frac{Q_c}{M_D \times Cp} \right) \tag{3.28}$$

where T_n is the temperature of the nth stage, T_{n+1} is the temperature at the $(n+1)$th stage, C_p is the specific heat of the liquid distillate and other notations have usual meanings.

The reactive stage temperature can be written as follows:

$$\frac{dT_n}{dt} = \frac{\left[\begin{array}{c} (L_{n-1}x_{n-1})(T_{n-1} - T_n) + (V_{n+1}y_{n+1})(T_{n+1} - T_n) \\ -(L_n x_n)(T_{n-1} - T_n) + (V_n y_n)(T_{n+1} - T_n) + (Fx_{feed})(fT_f - T_f) \end{array} \right]}{M_f} + \frac{V_R}{n} \sum R_{n,i} T_R \tag{3.29}$$

where T_R is the temperature of the reactive stage, T_f is the temperature of the feed, and other notations have usual meanings.

For reboiler, the equation written is as follows:

$$\frac{dT_n}{dt} = \frac{(L_{n-1}x_{n-1} - Bx_B)(T_{n-1} - T_n)}{M_B} + \frac{Q_B}{M_B \times Cp} \tag{3.30}$$

A MATLAB® (2013b) program has been developed for solving the above balance equations, rate equations, and temperature equations using ODE45 solver. ODE45 is an equation solver for solving stiff differential and algebraic equations [11].

3.3.11 Vapor–liquid equilibrium

At equilibrium condition, the relation between vapor and liquid fractions is given as follows:

$$y_{i,p} = K_{i,p} x_{i,p} \tag{3.31}$$

where y_{ip} is the vapor fraction of more volatile components, x_{ip} is the liquid fraction of more volatile components, and K_{ip} is the vapor–liquid equilibrium constant or proportionality constant.

Generally, the interface model for the nonequilibrium model is used, in which it is assumed that at interphase the physical equilibrium is attained as per the above equation.

3.3.12 Molecular weight

Average molecular weight is considered in the calculation for weight of the mixed product stream which can be calculated as follows:

$$M_{lp} = \sum M_{ilp} \times x_{ilp} \tag{3.32}$$

$$M_{vp} = \sum M_{ivp} \times y_{ivp} \tag{3.33}$$

where M_{ilp} is the average molecular weight of liquid, M_{ivp} is the average molecular weight of vapor, M_{ilp} is the molecular weight of ith component in the liquid phase, and M_{ivp} is the molecular weight of component i in the vapor phase.

3.3.13 Catalyst loading

It was found that with an increase in catalyst loading results in an increase in the rate of reaction and fractional conversion of acetic acid because of an increase in the number of active sites, which again indicates that mass transfer resistances are absent, and the reaction is only controlled by intrinsic kinetics. At higher catalyst loading, the rate of mass transfer is excessively high and therefore there is no more increase in the rate.

A general kinetic expression for esterification reaction catalyzed by heterogeneous catalysts can be written using the Langmuir equation as follows:

For reaction $A + B \rightarrow M + N$

$$-r_A = k_f \times W_{cat} \frac{a_i A \times a_i B - \left(a_i M \times \frac{a_i N}{Ke}\right)}{(1 + KA \times a_i A + KB \times a_i B + KM \times a_i M + KN \times a_i N)^2} \tag{3.34}$$

where k_f is the forward rate constant, a_i is the activity of the components, k_e is the esterification reaction constant, K_i is the equilibrium adsorption constant for each component, and W_{cat} is the weight of the catalyst.

We can calculate the weight of catalyst needed for the reactive zone by the following equation:

$$W = F_A \int \left[dx_A / (-r_A)\right] \tag{3.35}$$

where W is the weight of the catalyst (kg), F_A is the initial molar flow rate of component A (mol/min), and $-r_A$ is the rate of reaction (mole/min kg cat) [5–11].

3.4 Modeling of membrane-assisted reactive divided wall (MRDW) column

Modeling of reactive divided wall distillation column involves the basic concept of distillation column carrying out the reaction in a zone in between rectifying and stripping zones with a split zone in the middle of column which is separated by a component-selective membrane. Thus, modeling can be represented by various balances for different zones of reactive distillation column and that used for membrane separating the column vertically into two sections. The modeling of reactive distillation for different zones was already explained above.

The following assumptions are considered:

1. Equilibrium sorption of a permeating component is assumed to be at the membrane surface.
2. Due to strong interaction between the solute and the membrane material, the sorption equilibrium cannot be described using Henry's law.
3. Hence, sorption equilibrium is described using a chemical potential.
4. Fick's law is applicable in which the chemical potential is taken as a driving force.
5. Steady-state conditions are assumed for the membrane unit [5–11].

3.4.1 Material balance

At steady state, input = output

$$F = F_R + F_P \tag{3.36}$$

where F is the molar feed flow rate in mol/h, F_R is the molar retentate flow rate in mol/h, and F_P is the molar permeate flow rate in mol/h.

3.4.2 Component balance

$$F\, x_F = F_R x_R + F_P x_P \tag{3.37}$$

where x_F is the concentration of TAEE in feed, x_P is the concentration of TAEE in permeate, and x_R is the concentration of TAEE in retentate.

Fractional recovery is given as follows:

$$\theta = F_P / F \tag{3.38}$$

Hence, the component balance equation reduces to

$$x_F = (1 - \theta)\, x_R + \theta\, x_P \tag{3.39}$$

The chemical potential can be written as follows:

$$\mu_i = \mu_{io} + RT \ln \alpha_i \tag{3.40}$$

At sorption equilibrium, it can be written as follows:

$$\alpha_i = \Upsilon_i x_i \tag{3.41}$$

where Y_i is the activity coefficient of the ith component and x_i is the mole fraction of the component at equilibrium.

3.4.3 Mass transfer balance

Fick's law equation is modified in terms of chemical potential as follows:

$$J_i = - C_i B_i \, d\mu_i / dz \tag{3.42}$$

where B_i is the proportionality coefficient, C_i is the concentration of component, J_i is the flux, and $d\mu_i / dz$ is the driving force in terms of chemical potential gradient [1–11].

3.5 Simulation software (Aspen Plus)

Design engineers are responsible for drafting any plant layout; finalize the material of construction and the operating conditions to yield the desired aim. For the existing installed plants, output or plant efficiency can be increased only by modifying the current operating condition or modifying the machinery. Simulation is done to study the changes of design variables on the output. Simulation is an intimation of a real-world plant and one can properly study about the behavior of the plant.

The process model flow sheet maps out the entire system. The flow sheet is a general term given to depict all the entering stream, heat streams, and the outlets. Apart from streams involved, other utilities like pump, reboiler, heater, condenser, and manipulators, can also be added to the flow sheet to make a real-world imitation. Each stream and utilities are properly tagged with their name and operating conditions, making the system very user friendly. Aspen Plus also offers the special feature of NEXT button by which one can directly move to the step which requires further information to be supplied by the user. Thus, this reduces the calculation time, and the probability of human error reduces to the great extent.

3.5.1 Chemical components

The process simulator provides the list of chemical compounds based on the property method selected. A simulator can take the compound based on the molecular formula or its chemical name. A user can also develop his/her new chemicals by giving some information like density, molecular weight, boiling point, and freezing point.

3.5.2 Operating conditions

Operating conditions refer to temperature, pressure, flow rate, flow temperature, viscosity, and so on. This data should be known for all streams and unit operation so that the proper output results can be calculated. The number of variables involved should be equal to the number of equations so that the file converges with a zero degree of freedom.

ASPEN PLUS^{TM} allows us to create our own process model, starting with the flow sheet, then specifying the chemical components and operating conditions. There are two environments provided in Aspen Plus like property and simulation environment. In property environment, a user must give the chemical compounds and reactions with proper stoichiometry, and the property method should be chosen which should properly predict the system's behavior based on the binary interaction. The process simulator executes all necessary calculations needed to solve the outcome of the system, hence predicting its behavior. Once the property environment information is given, the user must enter the simulation environment. In simulation environment, the user has to provide the stream and flow sheet. The user should attach all the accessories such as pumps and valves to have rigorous outcomes.

Aspen Plus has a built-in property databank for calculation of feasibility of process which requires knowledge of thermodynamic and reaction engineering. Any missing parameter can be estimated automatically by the simulator because of these in-built data bank and binary interaction parameters. Aspen Plus can interactively change specifications such as flow sheet configuration, operating conditions, and feed compositions, to run new cases and analyze process alternatives. Aspen Plus allows us to perform a wide range of tasks such as estimating and regressing physical properties, generating custom graphical and tabular output results, fitting plant data to simulation models, optimizing process, and interfacing results to spreadsheets.

There are various property methods available in Aspen Plus software, which must be selected carefully based upon the characteristics of reaction to be carried out such as the phase of reactant and operating conditions. Some of them are listed below.

3.5.3 Nonrandom two-liquid models (NRTL)

The *nonrandom two-liquid* model (*NRTL* equation) belongs to the local composition models and is the most widely used property method for the systems exhibiting binary interaction among the feed and the product. This model works on an activity coefficient model, which correlates the activity coefficients y_i of a compound i with its mole fractions x_i in the liquid phase concerned. It is used to calculate the phase equilibria and is mostly applied for distillations, reactive distillation, dividing wall columns, and so on. In the concept of NRTL, bulk concentration differs from the local concentration around the molecule because of the difference between the interaction energy of the central molecule with the molecules of its own kind U_{ii} and that with the molecules of the other kind U_{ij}.

This energy difference introduces a nonrandomness which is also termed as entropy in chemical engineering thermodynamics. These local composition models are not thermodynamically consistent due to the assumption that the local composition around molecule i is independent of the local composition around molecule j. This assumption is not true, as was shown by Flemmer in 1976. Other methods available in this category are UNIQUAC, UNIFAC, and so on.

The parameters α_{12} and α_{21} are the so-called nonrandomness parameters, for which usually α_{12} is set equal to α_{21}. For a liquid in which the local distribution is random around the center molecule, the parameter $\alpha_{12} = 0$. In that case, the equations reduce to the one-parameter Margules activity model:

$$\ln \gamma_1 = x_2^2 [T_{21} + T_{12}] = Ax_2^2$$

$$\ln \gamma_2 = x_1^2 [T_{12} + T_{21}] = Ax_1^2$$

For aqueous system, generally α_{12} is set to 0.2, 0.3, or 0.48. The high value reflects the ordered structure caused by hydrogen bonds. However, in the description of liquid–liquid equilibria, the nonrandomness parameter is set to 0.2 to avoid wrong liquid–liquid description. In general, NRTL offers more flexibility in the description of phase equilibria than other activity models due to the extra nonrandomness parameters. However, in practice, this flexibility is reduced to avoid wrong equilibrium description outside the range of regressed data.

The limiting activity coefficients, the activity coefficients at infinite dilution, are calculated by

$$\ln \gamma_1^\infty = [T_{21} + T_{12\,exp}(-\alpha_{12}T_{12})]$$

$$\ln \gamma_1^\infty = [T_{21} + T_{12\,exp}(-\alpha_{12}T_{12})]$$

The expressions show that $\alpha_{12} = 0$ as the limiting activity coefficients are equal. This situation occurs for molecules of equal size, but of different polarities. It also shows, since three parameters are available, that multiple sets of solutions are possible.

3.5.4 Optimization using sensitivity analysis tool of Aspen Plus

Optimization of a process flow sheet is done to evaluate the value of input design variables that will provide the highest purity of products. This is done using revamping based on rigorous simulation and optimization in Aspen Plus process simulator.

Moreover, the sensitivity model analysis tool of Aspen Plus simulator is used to get an optimized value of a process variable within the range provided by the user. In the sensitivity analysis, a manipulated variable is defined, which will vary in a predefined interval under certain steps of increments and will show the values of design variable at each increment. The highest value of a design variable will be the optimized value corresponding to the manipulated variable. The optimized values of input design variables as obtained from revamping are further validated using sensitivity analysis under the model analysis tool of Aspen process simulator [12].

Table 3.1: Summary of Aspen property methods [12].

Abbreviation	Equation of state	
Peng–Robinson-based methods		
PENG-ROB	Peng–Robinson	
RKSWS	Redlich–Kwong–Soave with Wong–Sandler mixing rules	
RKSMHV2	Redlich–Kwong–Soave with modified Huron–Vidal mixing rules	
RK-ASPEN	Redlich–Kwong–ASPEN	
RK-SOAVE	Redlich–Kwong–Soave	
RKS-BM	Redlich–Kwong–Soave with Boston–Mathias alpha function	
Activity coefficient methods		
Abbreviation	**Liquid activity coefficient**	**Vapor fugacity coefficient**
NRTL-based methods		
ELECNRTL	Electrolyte NRTL	Redlich–Kwong
ENRTL-HF	Electrolyte NRTL	HF hexamerization model
ENRTL-HG	Electrolyte NRTL	Redlich–Kwong
NRTL	NRTL	Ideal gas
NRTL-HOC	NRTL	Hayden–O'Connell
NRTL-NTH	NRTL	Nothnagel
NRTL-RK	NRTL	Redlich–Kwong
NRTL-2	NRTL (using dataset 2)	Ideal gas
UNIFAC-based methods		
UNIFAC	UNIFAC	Redlich–Kwong
UNIF-DMD	Dortmund-modified UNIFAC	Redlich–Kwong–Soave
UNIF-HOC	UNIFAC	Hayden–O'Connell

Table 3.1 (continued)

Activity coefficient methods		
Abbreviation	**Liquid activity coefficient**	**Vapor fugacity coefficient**
UNIF-LBY	Lyngby-modified UNIFAC	Ideal gas
UNIF-LL	UNIFAC for liquid–liquid systems	Redlich–Kwong
UNIQUAC-based methods		
UNIQUAC	UNIQUAC	Ideal gas
UNIQ-HOC	UNIQUAC	Hayden–O'Connell
UNIQ-NTH	UNIQUAC	Nothnagel
UNIQ-RK	UNIQUAC	Redlich–Kwong
UNIQ-2	UNIQUAC (using dataset 2)	Ideal gas
WILSON-based methods		
WILSON	Wilson	Ideal gas
WILS-HOC	Wilson	Hayden–O'Connell
WILS-NTH	Wilson	Nothnagel
WILS-RK	Wilson	Redlich–Kwong
WILS-2	Wilson	Ideal gas
WILS-HF	Wilson	HF hexamerization model

References

[1] M.A. Khan, G. Adewuyi Yusuf, Techno-economic, modeling and optimization of catalytic reactive distillation for the esterification reactions in bio-oil upgradation, Chemical Engineering Research & Design: Transactions of the Institution of Chemical Engineers Part A, 148, 2019, 86–101.

[2] T. Lukács, C. Stéger, E. Rév, M. Meyer, Z. Lelkes, Feasibility of batch reactive distillation with equilibrium-limited consecutive reactions in rectifier, stripper, or middle-vessel column, International Journal of Chemical Engineering, 2011, 1–16.

[3] E.S. Lopez-Saucedo, I.E. Grossmann, J.G. Segovia-Hernandez, S. Hernández, Rigorous modeling, simulation and optimization of a conventional and nonconventional batch reactive distillation column: A comparative study of dynamic optimization approaches, Chemical Engineering Research & Design: Transactions of the Institution of Chemical Engineers Part A, 111, 2016, 83–99.

[4] E. Mityanina Olga, A. Samborskaya Marina, N. Sabiev Ayan, Mathematical modeling and steady states multiplicity analysis of methyl tert-butyl ether reactive distillation synthesis, Petroleum & Coal, 58, 4, 2016, 490–498.

[5]   A.G. Laptev, S.V. Karpeev, E.A. Lapteva, Modeling and modernization of tray towers for reactive distillation processes, Theoretical Foundations of Chemical Engineering, 52, 1, 2018, 1–10.
[6]   L. Jianwei, L. Zhigang, D.Z.L. Chengyue, C. Biaohua, Azeotropic Distillation: A Review of Mathematical Models, Separation & Purification Reviews, 34, Issue 1, 2005, 87–129.
[7]   F. Yun-Jin, L. Dao-Jun, A reactive distillation process for an azeotropic reaction system: transesterification of ethylene carbonate with methanol, Chemical Engineering Communications, 194, 12, 2007, 1608–1622.
[8]   O. Valderrama José, A. Faúndez Claudio, A. Toselli Luis, Advances on modeling and simulation of alcoholic distillation, Part 1: Thermodynamic modeling, Food & Bioproducts Processing: Transactions of the Institution of Chemical Engineers Part C, 90, 4, 2012, 819–831.
[9]   S. Sharma, V. Sakhre, S. Jain, Dynamic behavior of TAEE synthesis process in membrane assisted reactive distillation, Journal of Basic and Applied Engineering Research, ISSN: 2350-0255, JNU, Delhi-2014 20–24.
[10] V. Sakhre, S. Jain, V.S. Sapkal, D.P. Agarwal, Novel process integration of biodiesel blend in membrane assisted reactive divided wall (MRDW) column, Polish Journal of Chemical Technology, 18, 1, 2016, 105–112.
[11] V. Sakhre, S. Jain, V.S. Sapkal, Dev P. Agarwal, Modified Neural Network based cascaded control for product composition of Reactive Distillation, Polish Journal of Chemical Technology, 18, pp. 111–121, 2016.
[12] Fogler and Gurmen, Aspen PlusTM – Creating Flow Sheet, 2007, University of Michigan.

Chapter 4
Introduction to artificial neural networks

4.1 Why neural networks

With the adaptability of artificial intelligence (AI) and robotic applications, the neural networks (NNs), when compared with other computational intelligence algorithms, are more frequently used for solving problems. This is because of the learning ability of the NNs having the power of simulating the thinking capability of the human brain to solve the real-world complex problems. It has been observed that NNs have been used in the number of tasks for the purpose of pattern classification, robot or automobile vehicle navigation, object recognition, face detection, and so on. Therefore, in this chapter, we present the basic terminologies related to the concepts and theories related to the artificial NN [1].

4.1.1 History of neural networks

How a human brain works has been studied from thousands of years. And with the growth in technologies, it was advent to simulate it. In 1943, a paper on how NN may work was given by Warren McCulloch (a neurophysiologist) and Walter Pitts (a mathematician). The NN came into existence when it has been modeled with the electrical circuit. A book *Organization of Behaviour* written by Donald Hebb in 1949 reinforced the concepts of NN, while making out that neural pathways are strengthened each time that they are used. With advancement in computers in 1950s, it was possible to involve modeling how a human mind works. Nathanial Rochester (a researcher from IBM) tried to simulate an NN, failed in his first effort, but later he was succeeded. A boost to both AI and NNs in 1956 with the Dartmouth Summer Research Project has been seen letting to work with AI, compromising a lower level neural processing part of the brain. The work gets started very rapidly when John von Neumann gave the idea of imitating simple neuron functions by using telegraph relays or vacuum tubes. A brief history of the occurrence of different types of NNs has been given in Table 4.1. Research on NN and the related amount of fears to do work on NN needs to have some combined efforts to get the recognition. This period of enormous growth was noticed in 1981. Working with neurons to simulate how a brain works has been a hot topic for the researchers even in the present era with the new concepts of learning with such machine learning algorithms like support vector machine and extreme learning machine (ELM), making it a deep NN [1–2].

https://doi.org/10.1515/9783110656268-004

Table 4.1: Neuron description [1–18].

S. no.	Parts	Function	Components
1	Dendrites	Picks up the message from other neurons having finger-like projections from the cell body	Cell body acts as a message receiver
2	Nucleus	Heart of neuron which makes decision	
3	Cytoplasm	A liquid that surrounds the nucleus and passes information to the nucleus	
4	Schwann cells	Contains "myelin," a type of fat or protein that surrounds the axon as a sheath or insulator, making sure that the information does not leak out	Axon acts as a transmitter
5	Nodes of Ranvier	Helps ion polarizing a cell (approx. 70 mV), which is a gap between Schwann cells	
6	Dendrites	Connected with other neurons (known as "synapses" or "junction box" shown in Figure 4.1) to send a message, and if connected with the muscle or gland, it will simulate to do work	Axon terminal sends to others

4.2 Introduction to biological neurons

The architecture between the artificial NN and the biological neuron network has a similar structure. A biological neuron mainly consists of three elements known as dendrites, axon, and cell body or soma. The dendrites are responsible to link with the other neurons acting as a connection bridge that receives signals from the other neurons. These signals are then transmitted to the soma or cell body through the axon in the form of electric impulses or chemical processes. The incoming signals are then summed up in the cell body until an enough signal that is capable to fire up the cell has been received, through which the signal generated is transmitted to the other neuron connected through the dendrites. Figure 4.1 shows a neuron and Table 4.1 shows the neuron parts, description, and their working.

Some of the features that resemble the artificial network with biological network are as follows:

a. The processing elements are responsible for summation of the weighted inputs.
b. The received signal is modified with the help of weight associated with the synapses.
c. The output generated from the neuron or node is transmitted to the other neuron or nodes.
d. The learning procedure increases with the experience.
e. The process of information processing is local.

f.   Both have distributed memory, so the failure of one neuron or node does not affect the output.
g.   Both have the property of fault tolerance [1–3].

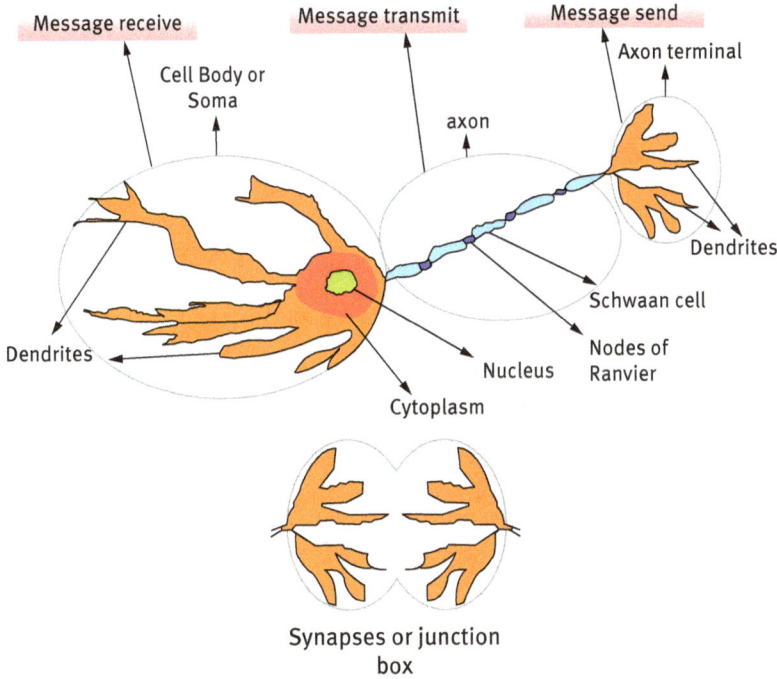

Figure 4.1: A neuron and synapses.

4.3 Introduction to artificial neural networks

NNs are the types of parallel computation inspired by neurons and their adaptive connections. An NN is used to solve practical problems by using a novel learning algorithm inspired by the brain. NN is an algorithm that built predictive models by learning the patterns in the historical data. They are made up of small, interconnected processing elements called nodes, and each node processes a small part of the task. Like nervous system, nodes are termed as neurons. Neuron helps transmit the message to the brain in the form of electrical current called nerve impulse finally delivered to the brain or spinal cord. To understand the NN, it is good to understand how the nervous system works.

So, the taxonomy of NNs can be summarized as follows:
a.   The process of information processing is achieved with the help of simple elements known as neurons.

b. Connection links are used to pass the signal among neurons.
c. Every connection link is associated with a weight that is being multiplied with the signal that is transmitted between the neurons.
d. Every neuron in the network is associated with an activation function that is used to optimize the given input signal to regulate the output signal [1–4].

4.4 Architecture of neural network

Architecture is a way in which nodes (neurons) are connected to each other. Most common NN is multilayer perceptron (MLP). Figure 4.2 shows an MLP, in which nodes are organized as layers. The architecture of NN looks like a computer system, that is, input = processing = output. A simple NN contains three layers: an input layer, a hidden layer, and an output layer. The independent variables are the input to the input layer. Then input to the hidden layer is $w_1u_1 + w_2u_2 + w_3u_3 + b$ (where w = weight and b = bias) total is processed by an activation function $f()$ and leaves the node as an output. The value of weight demonstrates the type of connection between the two, that is, strong or weak. The output from the output layer is y predicted. The NN then compares it with y actual (a dependent variable) if these do not match. It adjusts all the weights in the network and repeats the process until an accurate prediction is achieved for most of the observations and once this is achieved, we are left with an NN model that can be applied to a new set of data to provide predictions.

In mathematical terms, a neuron k is given by the following equations:

$$u_k = \sum_{j=1}^{m} w_{kj} x_j \tag{4.1}$$

$$y_k = \varphi(u_k + b_k) \tag{4.2}$$

where $x_1, x_2, \ldots, x_m$ are the set of input signals; $w_{k1}, w_{k2}, \ldots, w_{km}$ are the set of synaptic weights of k neuron; u_k is the linear combination output due to the input signals; b_k is the bias; $\varphi(\cdot)$ is the activation function; and y_k is the output signal of the neuron.

The use of bias is to achieve the affine transformation with the help of the equation

$$v_k = u_k + b_k \tag{4.3}$$

According to Bhattacharyya and Maulik in 2016, the working dynamics of the NN model can be understood by a simple matrix–vector multiplier given by the following mathematical equations.

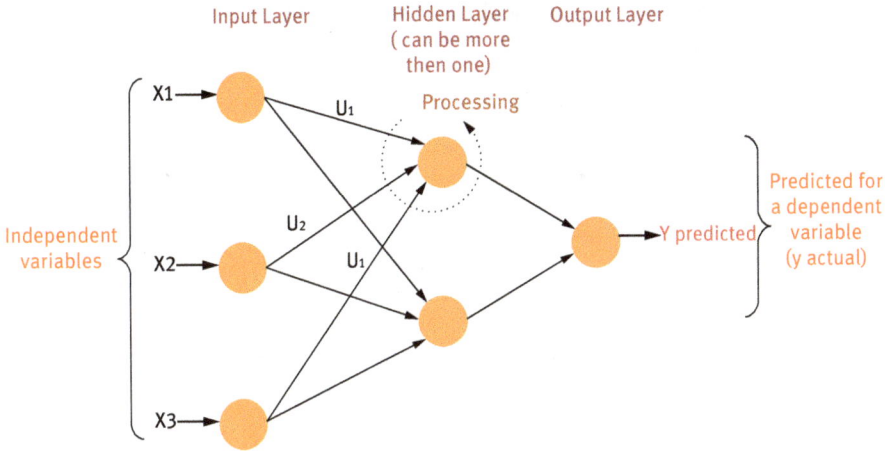

Figure 4.2: A multilayer perceptron (MLP).

Let n $\{a_1, a, \ldots, a_n\}$ be the set of inputs coming at the receiving neuron from n previous neurons, which exists in the vector space S_n, where $M_{b \times a} = \{m_{ij}, i = 1, 2, \ldots, a; j = 1, 2, \ldots, b\}$ is the interconnection weight matrix from b of such a receiving neuron to a preceding neuron. Then the summation approach in a neuron simplifies to a matrix–vector multiplication given by

$$
H = \begin{bmatrix} m_{11} & m_{12} & m_{13} & . & . & m_{1a} \\ m_{21} & m_{22} & m_{23} & . & . & m_{2a} \\ m_{31} & m_{32} & m_{33} & . & . & m_{3a} \\ . & . & . & . & . & . \\ . & . & . & . & . & . \\ m_{b1} & m_{b2} & m_{b3} & . & . & m_{ba} \end{bmatrix} \times \begin{bmatrix} q_1 \\ q_2 \\ q_3 \\ . \\ . \\ q_a \end{bmatrix} \tag{4.4}
$$

which is simply a vector given by

$$
H = \begin{bmatrix} h_1 \\ h_2 \\ h_3 \\ . \\ . \\ h_b \end{bmatrix} \tag{4.5}
$$

Thus, each jth $(j = 1, 2, \ldots, b)$ neuron is fed with an element of the vector $[h_1, h_2, \ldots, h_b]^T$ resulting out of the summing mechanism. As part of the processing task, each such neuron then applies the corresponding activation/transfer function (f)

to the elements of this vector to yield the corresponding output response O_j, $j = 1$, $2, \ldots, b$, as follows:

$$O = f \begin{bmatrix} h_1 \\ h_2 \\ h_3 \\ . \\ . \\ . \\ h_b \end{bmatrix} \tag{4.6}$$

Thus, given a set of inputs, the interconnection weight matrix $M_{b \times a} = \{m_{ij}, i = 1, 2, \ldots, a; j = 1, 2, \ldots, b\}$ along with the characteristic activation function f decides the output responses of the neurons, and thus the NN operates as a simple matrix–vector multiplier [1–5].

4.5 Difference between computers and neural networks

a. Computers more often use the set of instructions also known as algorithms or programs to solve some problem, but NNs differ in this process, that is, they learn from examples and then solve problems.
b. The operating capability of NN is different from conventional computers, that is, the network itself learns from the set of data that is being available for the purpose of training the model. If the data is inappropriate, the network may result in disaster. Whereas computers are programmed to perform certain tasks only, hence, the probability of error decreases.
c. In computers, the problem that has to be solved has been processed through the pipeline in which the user first writes his/her problem in any high-level language for his/her understanding, and then the computer interprets it and changes to low-level or machine-level language for its understanding. So, the point of error is generally due to the problem in the software content or hardware part. But in NN, the entire process depends on the quality of the data used for the purpose of training.
d. If utilized properly, the NNs can produce amazing results [1–6].

4.6 Properties of neural network

a. Learning of NN is accomplished mostly by backpropagation algorithm.
b. While designing NN, the following should be kept in mind:
    i. It should be good for things that the brain are at, for example, vision.
    ii. It should be bad for things that the brain are at, for example, 987*567.

c.   NNs are very slow learners.
d.   NN performs complex computation task.
e.   NN is not efficient as the brain but somewhat closer to it.
f.   The efficiency depends on the historical data used for the training.
g.   NNs can be a software or hardware.
h.   NNs are much intended toward the problems related to pattern recognition, detection, and classification [5–8].

4.7 Important parameters for neural network

a.   Epoch or number of cycles: it is the process in which the training samples have been processed and error has been calculated. The network is then back-propagated with the adjustment in the weights.
b.   Number of nodes and hidden layers: both the number of hidden layers and number of nodes in the hidden layer influence the performance of the network. So, while designing a network, these parameters should be carefully chosen. Otherwise, the network suffers the problem of overfitting or underfitting.
c.   The size and exactness of training samples: enough data with its correctness should be present to the network for its effective training and getting desired results.
d.   Learning rate: it is the most important factor that decides the learning capability of the NN, that is, how fast the network learns.
e.   Learning algorithm: the choice of learning algorithm is another most important factor that influences the working of the network. If a proper learning mechanism has not been adopted, then the network may result in a higher cost and time efficiency.
f.   Selection of the network: there are a number of NN models available, so a proper and suitable selection must be made depending upon the type and specificity of the problem [1–18].

4.8 Advantages and disadvantages of neural network

Advantages
a.   NN provides a solution for the problem that cannot be solved by using linear algorithms.
b.   An NN learns by its own and therefore it does not need programmers.
c.   Due to its parallel working scenario, if a node fails, the network can continue its working.
d.   NN can generalize.
e.   It can work with insufficient data also.

Disadvantages

a. NN is termed as a black box. So, the actual processing cannot be known.
b. Optimizing and adjusting the parameters is a difficult job to accomplish.
c. The processing time increases with increase in the size of data, number of layers, and number of nodes.
d. They need training to perform any task.
e. When the number of layers is increased, it is not easy to handle the complexity of the architecture.

4.9 Applications of neural network

Because of the learning mechanism and ability to solve complex problems, the NN has been used in several applications in different fields such as engineering, biomedical, defense, medical, and research.

a. Among all the computational intelligence, NN is widely used for the purpose of pattern recognition, for example, recognition of handwritten letters.
b. Computer vision applications like face detection, object extraction or segmentation, and pedestrian detection.
c. Machine intelligence like sorting and grading of crops.
d. Robotics and automation like navigation and control for a self-driven car.
e. Financial data analysis like stock market prediction.
f. Data processing and compression.
g. Game playing [1–18].

4.10 Learning algorithms for neural network

The learning is the fundamental task for NN. The output accuracy of the network depended completely on how a model has been trained using a learning mechanism. Basically, the network learning process has been categorized among three types as shown in Figure 4.3 supervised learning is the most widely used procedure for training the network. The network is trained with all the available samples in this process. The output of the network has been checked for the desired output, and the error is calculated if the desired results are not obtained. Then the network parameters have been adjusted according to the calculated error, and further the training process is accomplished until the desired output has been generated. In reinforced learning, the desired output is not available to the network, but instead the correctness of the output has been given. So, the network is trained until the correct output is generated. And in unsupervised learning, no desired or expected result has been known to the network. The network learns by its own from the features available from the input patterns and generates the best possible results [5–11].

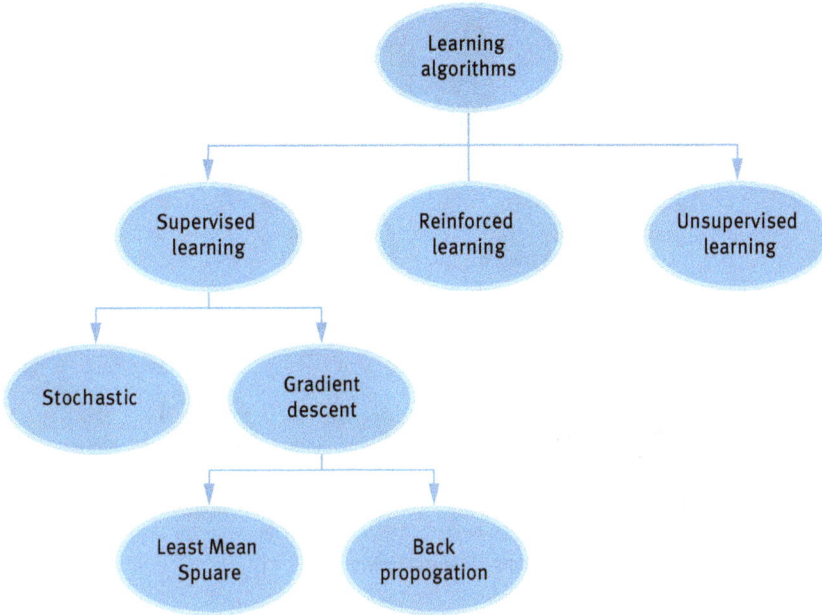

Figure 4.3: Taxonomy of learning algorithms of neural networks.

4.11 Classification and types of neural network

The NN is basically divided among two types given in Figure 4.4:
(i) Feedforward NN (FFNN) and
(ii) feedback/recurrent NN (RNN).

Further, we have tried to present some of the basic NN with their models and description as given below.

4.11.1 Perceptron

Perceptron was introduced by Frank Rosenblatt in year 1958. It is a computational model of the retina of the eye, and hence, it is named as "Perceptron." A single layer or multilayer feedforward network was best understood and extensively studied. Perceptron is also known as a binary classifier. However, the network can obtain weights only for linearly separable tasks. Perceptron adopts the supervised type of learning for the purpose of pattern classification. A simple perceptron model is given in Figure 4.5.

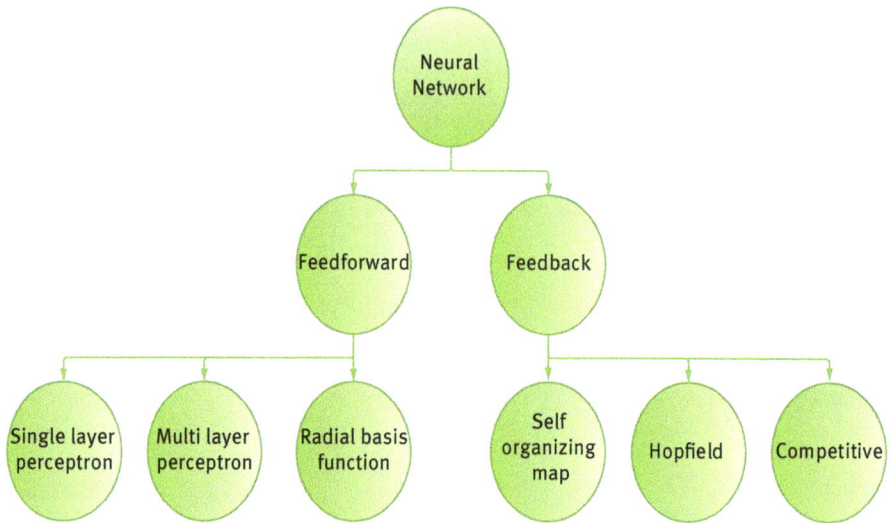

Figure 4.4: Taxonomy of learning algorithms of neural networks.

4.11.2 ADALINE (adaptive linear neuron element)

ADALINE (adaptive linear neuron element) network was introduced by Bernard Widrow in 1960. It was grounded on McCulloch–Pitts neuron. Like Perceptron, ADALINE is also a single-layer NN. It is based on the mechanism to lower down the mean square error after each iteration of the network. It is based on the supervised learning algorithm and is like Perceptron. It is used in all high-speed modems and telephone switching systems. A simple ADALINE network is given in Figure 4.6. The difference between the ADALINE and PERCEPTRON is that the ADALINE is known for minimizing continuous error, whereas the PERCEPTRON is known for minimizing the binary error.

4.11.3 MADALINE (many adaptive linear neuron element)

MADALINE (many adaptive linear neuron element) network was introduced by Bernard Widrow in 1960 and 1988. It is connected by a combination of ADALINE networks spread across multiple layers with adjustable weights. The network employs a supervised learning rule called MADALINE adaptation rule (MR) based on "minimal disturbance principle." The network is like MLP, which can solve the problem of nonlinear function classification. Like ADALINE, MADALINE has constant values for weights interconnecting nodes and bias value with "1." The multilayer network is shown in Figure 4.7.

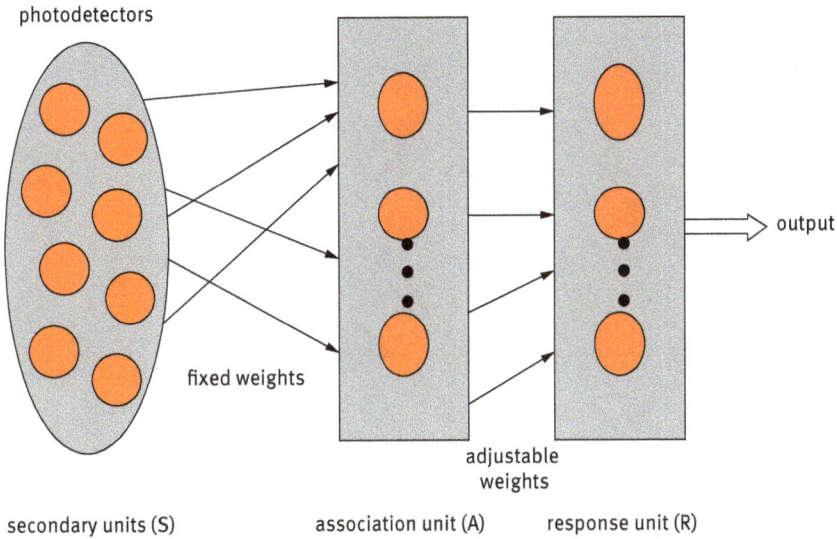

Figure 4.5: Perceptron network.

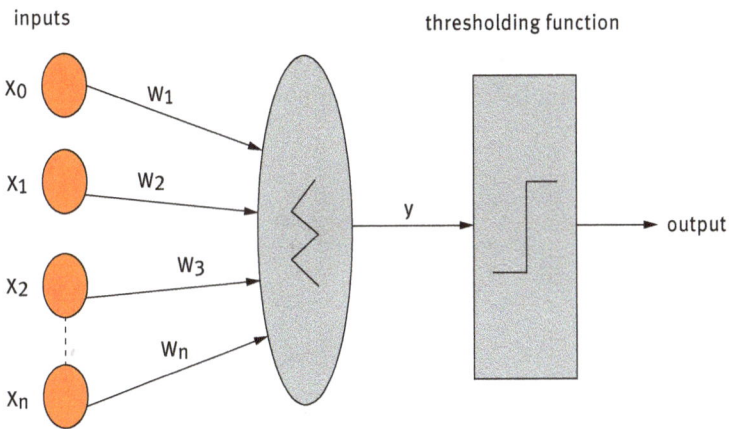

Figure 4.6: ADALINE network.

4.11.4 Feedforward neural network (FFNN)

FFNN is the simplest and most common network. In these networks, the information flows only in one direction without forming a loop. If there is more than one hidden layer, we call them deep NN, as shown in Figure 4.8. Generally, the network constitutes of the three layers known as the input, the hidden, and the output layers. All the layers relate to the help of connecting links. The input signal from

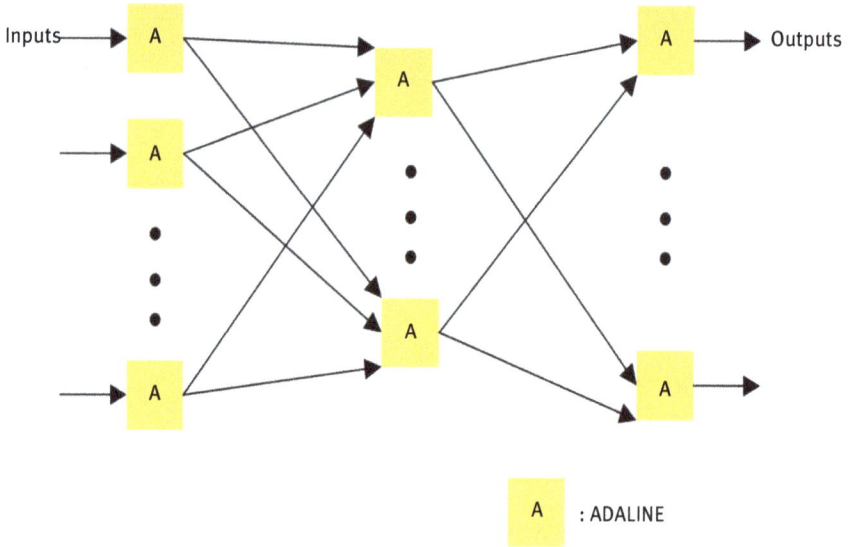

Figure 4.7: MADALINE network.

the input layer has been processed in the hidden layer with the help of weights associated with the links, and the output layer is responsible for presenting the results. They are helpful in computation of a series of transformations that change the similarities among cases.

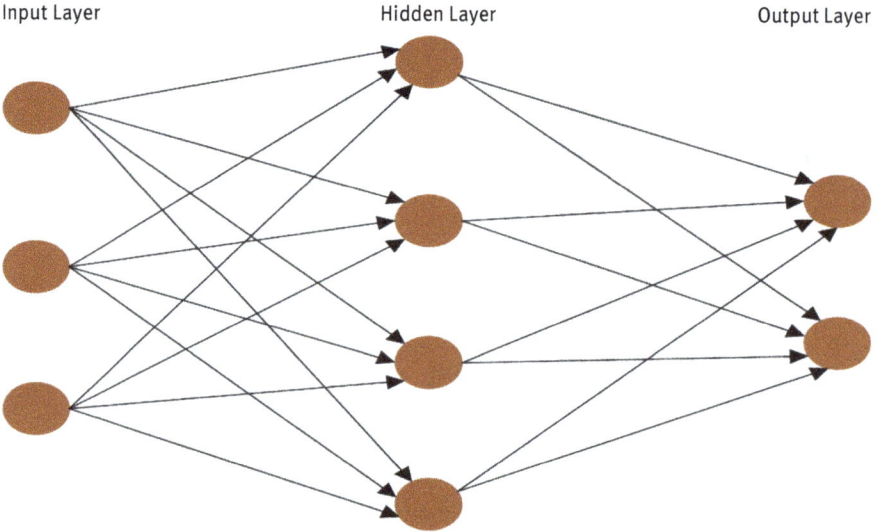

Figure 4.8: Feedforward neural network.

4.11.5 Adaptive resonance theory (ART)

Adaptive resonance theory (ART) was introduced by Carpenter, Grossberg, and others in 1980. The network employs a new principle of self-organization called ART based on competitive learning or unsupervised learning. The term "resonance" here means the resonant state of the network in which the category prototype vector matches close enough to the current input vector. Figure 4.9 shows a simple ART model. In general, the ART model is classified among two types: ART1 (used for clustering binary vectors) and ART2 (used for clustering continuous vectors). The drawback of the network structure is its complexity [5–17].

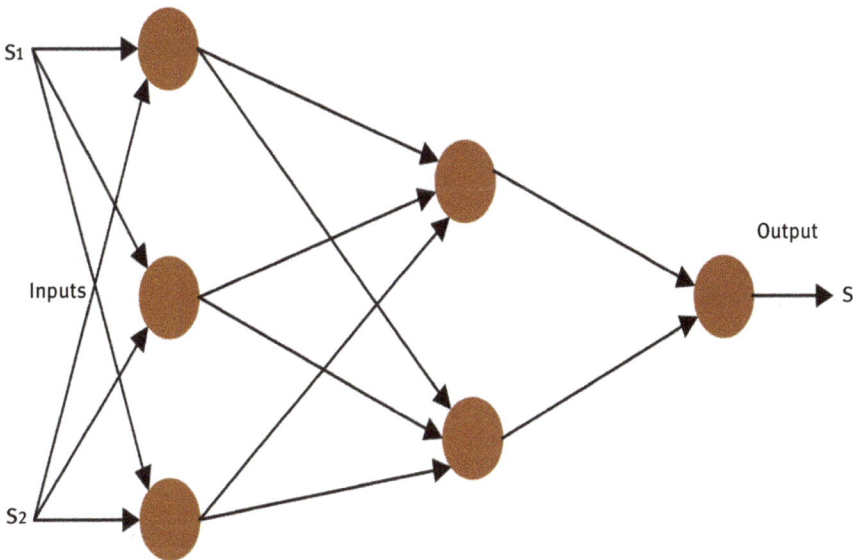

Figure 4.9: Adoptive resonance theory (ART).

4.11.6 Backpropagation network (BPN)

Backpropagation network (BPN) was proposed by Rumelhart, Hinton, Williams in 1985, Werbos in 1974, and Parker in 1985. It is a learning algorithm applicable to any feedforward network architecture. The error of conventional network has been calculated and weights are updated with the help of backpropagation. With the help of this method the networks can adapt to the desired results. A simple network architecture of the feedforward network with backpropagation strategy is given in Figure 4.10. Slow rate of convergence and local minima problem are its weakness.

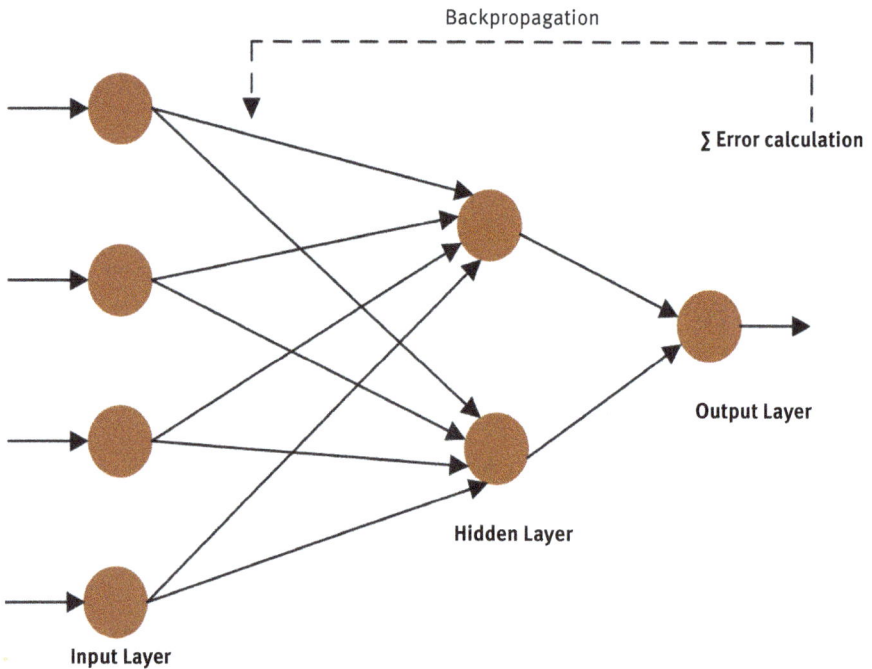

Figure 4.10: Backpropagation network.

4.11.7 Recurrent neural network (RNN)

In this network, information flows in a cycle and it can remember it for a long time. More powerful than FFNN, sometimes you get back to where you have started as if they have directed cycles. These networks are designed to identify the data sequential attributes and use the pattern to predict the result. Have a complicated diagram and it is not easy to train them having a property of biological realization. This network is mostly used for the task of speech recognition and natural language processing. The important characteristics of the network are that they are robust and powerful among family of NN and therefore used in the deep learning. RNN is shown in Figure 4.11.

4.11.8 Boltzmann and Cauchy machines

These were introduced by Hinton and Sejnowski in 1983, 1985, and 1986, and Szu H. and E. Hartley in 1987. These are stochastic networks whose states are governed by the Boltzmann distribution/Cauchy distribution. The heavy computational load is

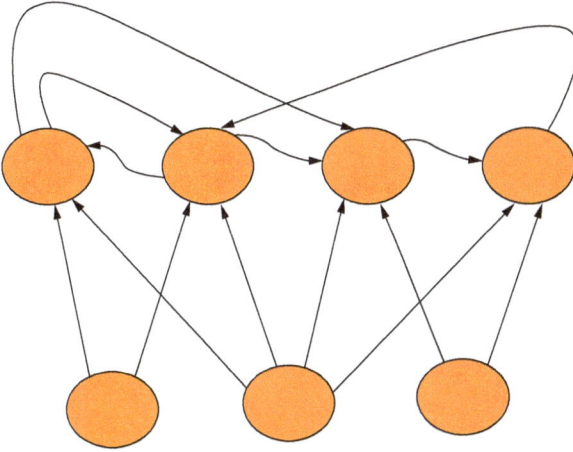

Figure 4.11: Recurrent neural network.

its drawback. The main objective of the Boltzmann machine is to optimize weights of the network for getting the desired results. A simple model for the Boltzmann machine is given in Figure 4.12, having the recurrent structure. The neurons in the machine are in either of the two states, that is, free state or frozen state.

4.11.9 Counterpropagation network

Counterpropagation network was proposed by Robert Hecht Nielsen in 1987. The proposed network combines the features of two networks. The network belongs to the category of self-organization networks and functions as statistically optimal self-programming lookup table. The weight adjustments between the layers follow Kohonen's unsupervised learning rule and Grossberg's supervised learning rule [5–12]. Counterpropagation network is shown in Figure 4.13.

4.11.10 Hopfield network

This network was introduced by John Hopfield in the year 1982. It is a single-layer recurrent network which makes use of Hebbian learning or gradient descent learning. The network also has associate memory with it. The important characteristic of this network is that in this all the nodes act as an input and output. The number of feedback loops are also like the number of neurons in the network. The weights are symmetric and no connection to the node itself is allowed. Also, the network has no hidden layer as shown in Figure 4.14.

visible neurons   hidden neurons

1

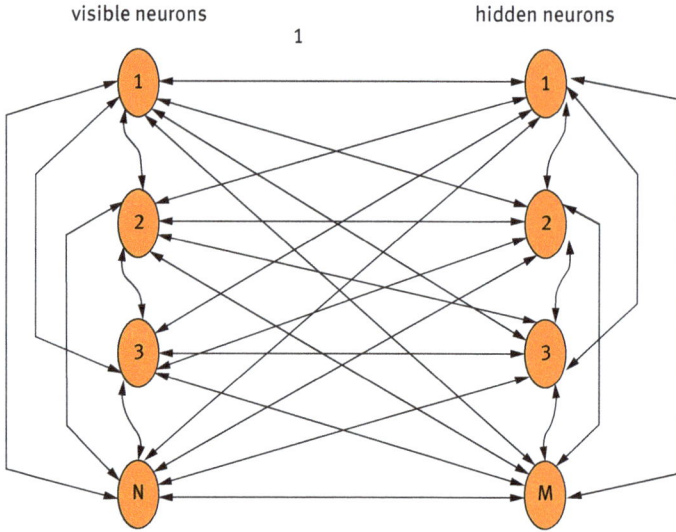

Figure 4.12: Boltzmann and Cauchy machines.

3 hidden layers

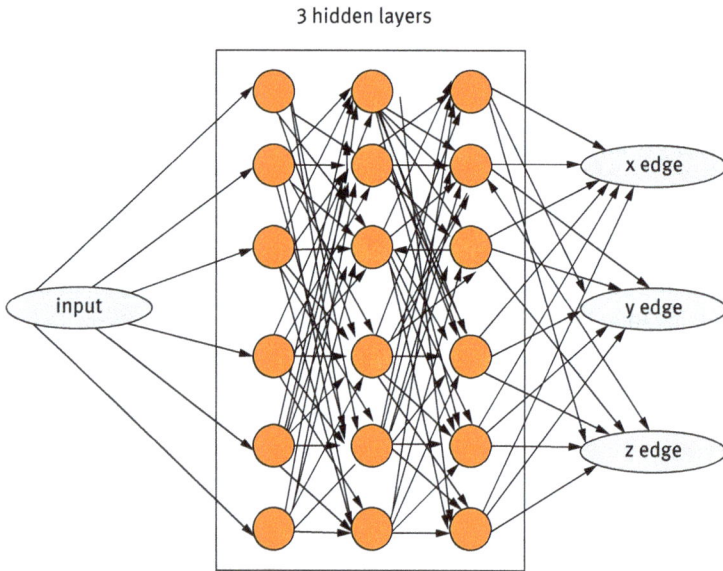

Figure 4.13: Counterpropagation network.

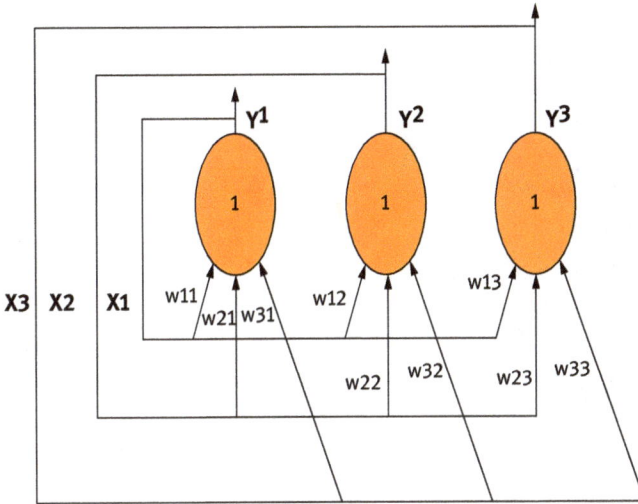

Figure 4.14: Hopfield network.

4.11.11 Self-organizing feature map (SOFM)

Self-organizing feature map (SOFM) networks were introduced by Kohonen in 1982. The network is a simplified model of the feature to localized region mapping of a brain. It is a self-organizing network that employs competitive learning or unsupervised learning. This is the way it differs from the other conventional networks that use backpropagation learning for updating these weights of the network. These networks can generate the output of visual representation with rectangular or hexagonal grid. Figure 4.15 illustrates the SOFM architecture [6–17].

4.11.12 Pulse-coupled neural network (PCNN)

Pulse-coupled NN (PCNN) was introduced by Eckhorn in 1990, inspired by biological studies on the method underlying the visual cortex of the small mammals. It has been widely used in image recognition, object extraction, edge detection, texture analysis, and image processing. PCNN is a single-layer, 2-D, and laterally connected network of integrate-and-fire neurons. A simple PCNN model is given in Figure 4.16.

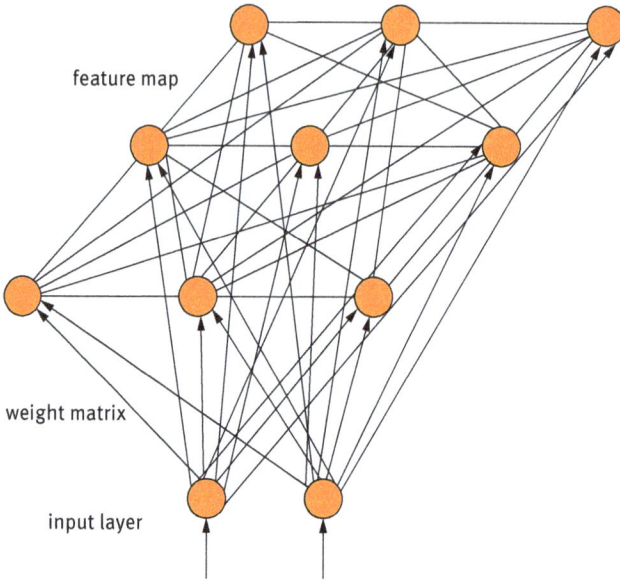

Figure 4.15: Self-organizing feature map.

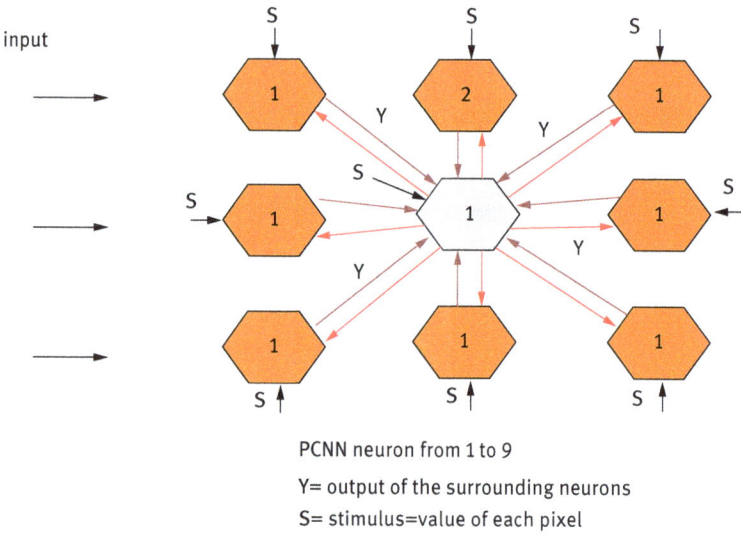

PCNN neuron from 1 to 9
Y= output of the surrounding neurons
S= stimulus=value of each pixel

Figure 4.16: Pulse-coupled neural network.

4.11.13 Bidirectional associative memory

This was introduced by Bart Kosko in 1988. This is a two-layer recurrent, hetero-associative network that can store pattern in pairs and retrieve them. It behaves as content addressable memories.

4.11.14 Neocognitron

This was introduced by Kunihiko Fukushima in 1982. This is a hybrid hierarchical multilayer feedback/feedforward network which closely models a human vision system, and the network employs either supervised or unsupervised learning rules [5–18].

4.12 Latest developments to the family of neural networks

Whereas with the development in the technologies, some latest artificial NNs have been introduced to the family of NNs. In this section, we will focus on introducing some of them.

4.12.1 Generative adversarial networks

The concept of these kinds of networks has been introduced by Ian Goodfellow in the year 2014. These networks are designed in combination of the two NN architectures. More often, the combination of FFNN and convolutional NN exists. In this combination, the one model is being treated as generative and the other one is discriminative. The discriminative one classifies the given image among natural or artificial one, and the generative has a task of generating images like the original data distribution [7–18].

4.12.2 Neuro-fuzzy system (NFS)

Neuro-fuzzy system is based on the fuzzy system whose training is accomplished with the help of NN methodology. It resembles a three-layer NN that comprises input, hidden, and output layers. Where input layer represents input variables, hidden layer corresponds to the set of fuzzy rules, and output layer represents the output variables. More often the hybrid network is a fuzzy-based system only. Semantical properties of the underlying fuzzy system have been considered for the learning procedure in these networks. Generally, the hybrid model is represented in two forms shown in Figure 4.17.

Model 1: Fuzzy system neural network   Model 2: Neural network-fuzzy system

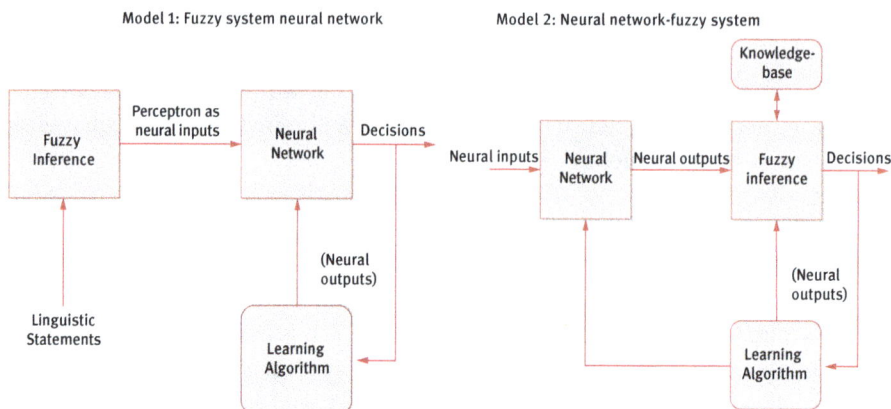

Figure 4.17: Neuro-fuzzy models.

The proposed models are having the learning capabilities of NNs and with the representational properties of the fuzzy system. One of the drawbacks of such systems is the rules defined using learning process that are not necessarily understandable [5–18].

4.12.3 Deep neural network (DNN)

DNN is the latest research topic among researchers and academicians. This network is known for its complex architecture and for solving the problem of classification with higher accuracy. Generally, the network has a layered structure comprising convolutional layers, max-pooling layer, and fully connected layers. This network is known for its classification accuracy. GoogleNet, AlexNet, and ResNet are some of the most widely used architectures among this category. Deep learning models are harder to train with their complex structure, but once they are trained, they produce high-quality results. They require a large volume of data to train. The application of deep learning lies from object detection to speech recognition [5–17]. A simple deep learning model is given in Figure 4.18. Some of the essential parameters of the network are as follows:
a) Number of epochs
b) Training algorithm
c) Learning rate
d) Weight decay
e) Activation function
f) Momentum
g) Dropout
h) Batch size
i) Number of convolutional
j) Number of pooling layers

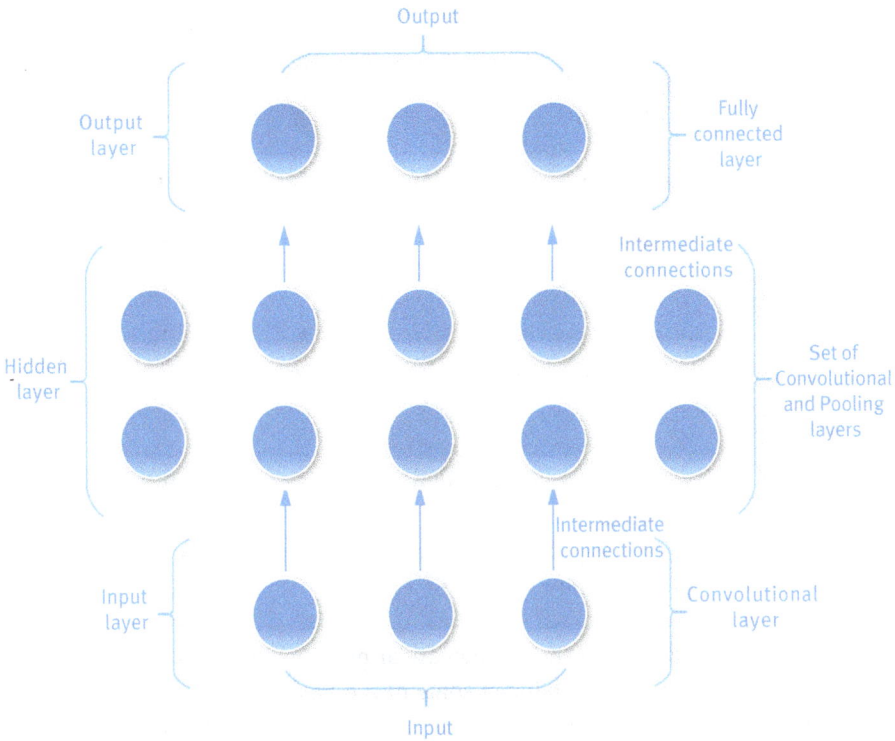

Figure 4.18: Deep learning model.

4.12.4 Extreme learning machine (ELM)

These networks are introduced for removing the drawbacks of conventional FFNNs. They overcome the slow gradient learning approach of the model and tuning of parameters iteratively. The proposed ELM is treated as machines that constitute the number of hidden layers, whose initial weights are assigned randomly. These machines adopt the concept of early perception model and random projection in solving complex problems. The concept of ELM lies on the theory of empirical risk minimizing theory, and its learning is accomplished in only one iteration unlike convolutional NN. It overcomes the problem of local minimization. It has been widely used in data analysis and image processing [5–17]. ELM is shown in Figure 4.19.

4.13 Summary
In this chapter, we have introduced the concept and theory related to the artificial NN family. This chapter describes all the basic terminologies related to the neuron and its working architecture. The mechanism behind the learning algorithms and

Input layer   Hidden layer   Output layer

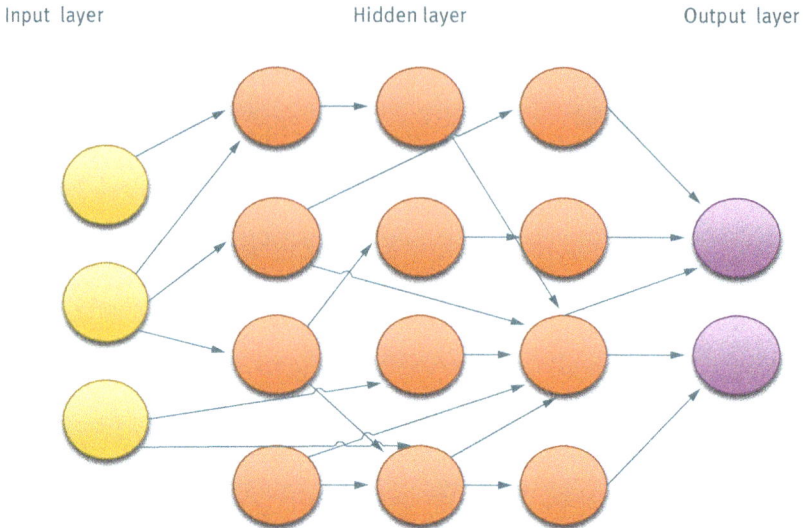

Figure 4.19: Extreme learning machine.

the architectural details with the mathematical proof have also been included in this chapter. We have also introduced some of the basic and latest networks that are being adopted to solve the real-life complex problems. Finally, we believe that this chapter presents a brief introduction to the NN and its terminologies and will help the academicians and researchers [18–20].

References

[1] Y. Bengio, A. Courville, P. Vincent, Representation learning: A review and new perspectives, IEEE Transactions on Pattern Analysis and Machine Intelligence, 35, 8, 2013, 1798–1828.
[2] J.J. Hopfield, Neural networks and physical systems with emergent collective computational abilities, Proceedings of the National Academy of Sciences of the United States of America, 79, 8, 1982, 2554–2558.
[3] N. Rochester, J.H. Holland, L.H. Habit, W.L. Duda, Tests on a cell assembly theory of the action of the brain, using a large digital computer, IRE Transactions on Information Theory, 2, 3, 1956, 80–93.
[4] F. Rosenblatt, The perceptron: A probabilistic model for information storage and organization in the brain, Psychological Review, 65, 6, 1958, 386–408.
[5] V. Sakhre, U.P. Singh, S. Jain, FCPN approach for uncertain nonlinear dynamical system with unknown disturbance, International Journal of Fuzzy Systems, 19, 4, 2017.
[6] U.P. Singh, S.S. Chouhan, S. Jain, S. Jain, Multilayer convolution neural network for the classification of mango leaves infected by anthracnose disease, IEEE Access, 2019.
[7] T. Cover, P. Hart, Nearest-neighbour pattern classification, IEEE Transactions on Information Theory, 13, 1, 1967, 21–27.

[8]   N.A. Campbell, J.B. Reece, Biologie. Spektrum, 2000, Akademischer Verlag.

[9]   G. Cybenko, Approximation by superpositions of a sigmoidal function, Mathematics of Control, Signals, and Systems, 2, 4, 1989, 303–314.

[10] U.P. Singh, S. Jain, Optimization of neural network for nonlinear discrete time system using modified quaternion firefly algorithm: case study of Indian currency exchange rate prediction, Soft Computing (Springer), 22, 8, 2018, 2667–2681.

[11] U.P. Singh, S. Jain, A.K. Tiwari, R.K. Singh, Gradient evolution based counter propagation network for approximation of noncanonical system, Soft Computing (Springer), 2018.

[12] R.O. Duda, P.E. Hart, D.G. Stork, Pattern Classification, 2001, Wiley, New York.

[13] J.J. Hopfield, Neural networks and physical systems with emergent collective computational abilities, Proceedings of the National Academy of Science, USA, 79, 1982, 2554–2558.

[14] R.S. Sutton, A.G. Barto, Reinforcement Learning: An Introduction, 1998, MIT Press, Cambridge, MA.

[15] U.P. Singh, S. Jain, Modified chaotic bat algorithm-based counter propagation neural network for uncertain nonlinear discrete time system, International Journal of Computational Intelligence and Applications, 15, 3, 2016.

[16] A. Scherbart, N. Goerke, Unsupervised system for discovering patterns in time-series, 2006.

[17] S. Agarwal, R.K. Singh, U.P. Singh, S. Jain, Biogeography particle swarm optimization based counter propagation network for sketch based face recognition, Multimedia Tools and Applications, 78, 8, 2019, 9801–9825.

[18] A.S. Weigend, N.A. Gershenfeld, Time Series Prediction, 1994, Addison Wesley.

[19] K.S. Narendra, K. Parthasarathy, Neural networks and dynamical systems, International Journal of Approximate Reasoning, 6, 1992, 109–131.

[20] A. Kroll, H. Schulte, Benchmark problems for nonlinear system identification and control using Soft Computing methods: Need and overview, Applied Soft Computing Journal, 25, 2014, 496–513.

Chapter 5
Application of artificial neural network

5.1 Background

Although the reactive distillation process is very effective in all types of manufacturing including the equilibrium-limited reaction and provides more effective simultaneous separations, it is highly nonlinear in nature as there are many interactive variables involved, which affects the overall conversion and product performance parameters. Linear control theory does not provide a good control action in complex and nonlinear systems and leads to instable controller performance. Moreover, when poorly known plant dynamic characteristics and unpredictable variations occur, the high-performance controllers are difficult to implement. Such type of problems can be overcome with the use of nonlinear control systems. Various types of nonlinear controllers for this purpose are robust controllers, adaptive controllers, neural network controllers, fuzzy logic controllers, and hybrid controllers. Complex, interacting, and nonlinear processes can be effectively identified and controlled using neural network, fuzzy logic, and the hybrid of neuro-fuzzy control techniques [1].

5.1.1 Artificial neural network (ANN)

Artificial neural networks (ANNs) are nonlinear information processing devices, which are built from interconnected elementary processing devices called neurons. It is inspired from the biological nervous system, such as the human brain, and process information. It is composed of a large number of highly interconnected neurons (processing elements) working to solve nonlinear problems. The signals are transmitted by means of connection links. The connected links process an associated weight, which is multiplied with the net input. The output signal is obtained by applying activations to the net input. ANNs have the potential for improving performance through dynamical learning. ANN can be considered as a large dimensional nonlinear dynamical system, which is defined by the set of first-order nonlinear differential equations.

ANN is used to learn patterns and relationship of the data. The data may be from the process industry and market research. ANN does not require any traditional coding approach to solve the problem. For example, to generate a model for having a predicted output of the reactive distillation column, ANN needs to be given only the raw data which is related to this problem. The data that is given might consist of past historical input data, temperature variant, pressure, flow rate, and other related variables. The ANN processes this information and produces an understanding of the factors affecting the output. The model can then be used for prediction of the output based on the key factors. The ANN works on the learning

https://doi.org/10.1515/9783110656268-005

rules based on the various algorithms used to learn the relationship in the data. The learning rules help the network to gain knowledge from the available data and apply the knowledge to assist a manager in taking an important decision to control the process. ANN is the simplification of real biological network of neurons. Most of the application uses the behavior of a single neuron as the basic computing neuron for explaining the processing operation of neuron information. The experimental studies on neuropsychology show that the response of a biological neuron appears to be random, and only by averaging much operation, it is possible to predict the results. Inspired by this operation, some researchers have developed neural structures based on the concept of neural population [1–5].

5.2 Adaptive linear neuron (ADALINE)

Single-layer neural network with adaptive learning systems/single-layer unit was first introduced by Bernard Widrow and Marcian E. Hoff in 1960 and is commonly known as adaptive linear neuron (ADALINE). It could be found in nearly every real-time noise filtering and its training is done using Widrow–Hoff rule or delta learning rule. The advantages of the delta learning rule over a perceptron network are its adaptability, that is, if the difference between the target and the actual output of the network are not small, then change the weight of network accordingly. In Adaline, the input–output relation is linear and the delta learning rule is based on the gradient-descent approach, which updates the weights between the connection to minimize the difference between the calculated output and target.

For the training of the Adaline, we consider the following notations: let us assume that the input and output patterns of the input layer are $I = (I_{I1}, I_{I2}, \ldots, I_{IM})$ and $I = (I_{O1}, I_{O2}, \ldots, I_{OM})$, where I_{Ii} and I_{Oi} are the input and output at the ith unit at the input layer, respectively. Similarly, $H = (H_{I1}, H_{I2}, \ldots, H_{IN})$ and $H = (H_{O1}, H_{O2}, \ldots, H_{ON})$, where H_{Ij} and H_{Oj} are the input and output at the jth unit at the hidden layer, respectively, also let $O = (O_{I1}, O_{I2}, \ldots, O_{IK})$ and $O = (O_{O1}, O_{O2}, \ldots, O_{OK})$, where O_{Ik} and O_{Ok} are the input and output at the kth unit at the output layer, respectively, and targets are denoted as $T = (T_{O1}, T_{O2}, \ldots, T_{OK})$, where $1 \leq i \leq M$, $1 \leq j \leq N$, and $1 \leq k \leq K$. Also v_{ij} denotes the connecting weight between the ith unit of the input layer to the jth unit of the hidden layer, and w_{jk} is the connecting weight between the jth hidden unit to the kth output unit as shown in Figure 5.1.

The delta learning rule for weight adjusting is given by $\Delta w_{jk} = \alpha(T_{Ok} - O_{Ok})I_{Ii}$. For the training set (I, T) with cardinality P, let us consider the cost function given as follows:

$$E = \frac{1}{P}\sum_{k=1}^{P}(T_{Ok} - O_{Ok})^2 \tag{5.1}$$

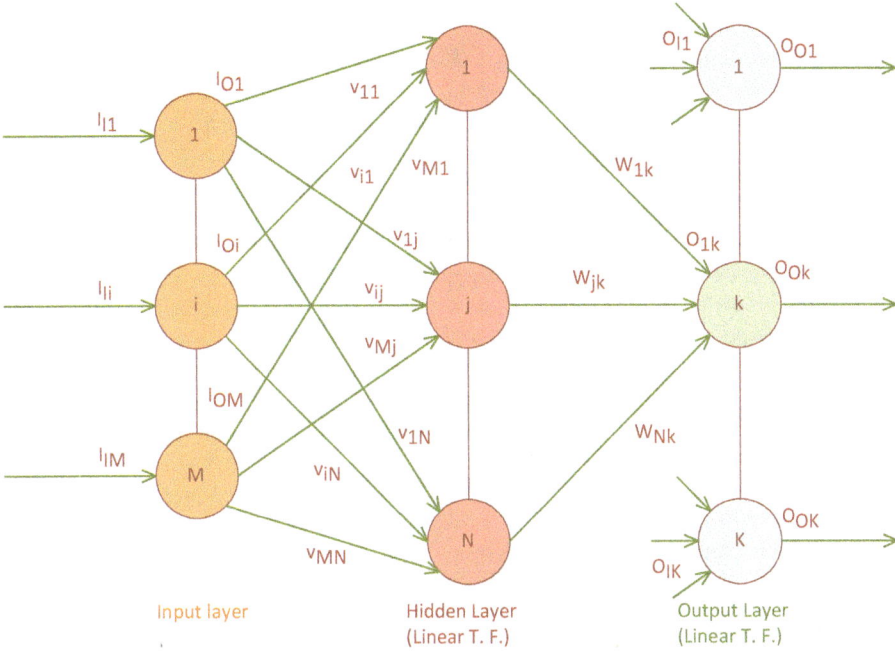

Figure 5.1: Adaline network.

$$E = \frac{1}{P}\sum_{k=1}^{P}(E_k)^2, \text{ where } E_k = T_{Ok} - O_{Ok}$$

To minimize the cost function, the incremental mode of training is performed in two steps.

Step I: To update the connecting weight between the hidden and output layers:

$$w_{jk}(\text{updated}) = w_{jk}(\text{previous}) + \Delta w_{jk} \tag{5.2}$$

where

$$\Delta w_{jk} = -\eta \frac{\partial E}{\partial w_{jk}} \tag{5.3}$$

Evaluate the gradient:

$$\frac{\partial E_k}{\partial w_{jk}} = \frac{\partial E_k}{\partial O_{Ok}}\frac{\partial O_{Ok}}{\partial O_{Ik}}\frac{\partial O_{Ik}}{\partial w_{jk}} \text{ (using chain rule at the } k\text{th neuron)} \tag{5.4}$$

where

$$\frac{\partial E_k}{\partial O_{Ok}} = -(T_{Ok} - O_{Ok}) \tag{5.5}$$

$$\frac{\partial O_{Ok}}{\partial O_{Ik}} = 1 \tag{5.6}$$

$$\frac{\partial O_{Ik}}{\partial w_{jk}} = H_{Oj} \tag{5.7}$$

Using eqs. (5.5)–(5.7), eq. (5.4) yields

$$\frac{\partial E}{\partial w_{jk}} = \frac{1}{P}\sum_{k=1}^{P}\frac{\partial}{\partial w_{jk}}(T_{Ok} - O_{Ok})^2$$

$$\frac{\partial E_k}{\partial w_{jk}} = -(T_{Ok} - O_{Ok})H_{Oj} \tag{5.8}$$

Step II: To update the connecting weight between input and hidden layers:

$$v_{ij}(\text{updated}) = v_{ij}(\text{previous}) + \Delta v_{ij} \tag{5.9}$$

where $\Delta v_{ij} = -\eta \left\{\dfrac{\partial E}{\partial v_{ij}}\right\}_{\text{avg}}$ and $\left\{\dfrac{\partial E}{\partial v_{ij}}\right\}_{\text{avg}} = \dfrac{1}{P}\sum_{k=1}^{P}\dfrac{\partial E_k}{\partial v_{ij}}$

$$\frac{\partial E_k}{\partial v_{ij}} = \frac{\partial E_k}{\partial O_{Ok}}\frac{\partial O_{Ok}}{\partial O_{Ik}}\frac{\partial O_{Ik}}{\partial H_{Oj}}\frac{\partial H_{Oj}}{\partial H_{Ij}}\frac{\partial H_{Ij}}{\partial v_{ij}} \quad \text{(using chain rule at the }k\text{th neuron)} \tag{5.10}$$

where

$$\frac{\partial E_k}{\partial O_{Ok}} = -(T_{Ok} - O_{Ok}) \tag{5.11}$$

$$\frac{\partial O_{Ok}}{\partial O_{Ik}} = 1 \tag{5.12}$$

$$\frac{\partial O_{Ik}}{\partial H_{Oj}} = w_{jk} \tag{5.13}$$

$$\frac{\partial H_{Oj}}{\partial H_{Ij}} = a_1 H_{Oj}(1 - H_{Oj}) \quad \text{(activation function is log – sigmoidal)} \tag{5.14}$$

$$\frac{\partial H_{Ij}}{\partial v_{ij}} = I_{Ij} \tag{5.15}$$

Using eqs. (5.11)–(5.15), eq. (5.10) yields

$$\frac{\partial E_k}{\partial v_{jk}} = -a_1(T_{Ok} - O_{Ok})H_{Oj}(1 - H_{Oj})w_{jk}I_{Ij} \tag{5.16}$$

5.3 Backpropagation neural network

Neural network is a network of nodes, and its variety depends on how the nodes are connected. One of the drawbacks of the Adaline network is the used linear activation function on hidden and output layers. To remove it, we apply that the learning algorithm is applied to multilayer feedforward neural network (MFNN) consisting of processing elements with nonlinear continuous differentiable activation functions. The networks associated with these functions and backpropagation learning are known as backpropagation neural network (BPNN) as shown in Figure 5.2 for the given set of input–output pair, and backpropagation method provides a procedure for changing weights. The concept of weight update methods is steepest gradient descent with differentiable activation functions. The general difficulties in the BPNN with multilayer are calculating the weight of hidden layer efficiently, which causes minimum output error. If the hidden layer increased, the learning of the BPNN becomes more complex. Also, the error is easily measured at the output layer but there is no such mechanism at the hidden layer.

The neurons of the BPNN are grouped as one input layer, hidden layers, and one output layer. The nodes in the input layer of the network supply respective elements which constitute the input signals applied to the computational nodes, that is, hidden layer. The output signals of these hidden layers are used as input to the output layer. Typically, the number of neurons in the input and output layers are known but the number of neurons in the hidden layer can be determined by the parametric study or the method of clustering []. For brevity, the network shown in Figure 5.2 is represented as $M–N–K$, that is, in BPNN with only one hidden layer containing M number of nodes in the input layer, N number of neurons in the hidden layer, and K number of neurons in the output layer. The neural network shown in Figure 5.2 is called as a fully connected network in the sense that each neuron in every layer of the network relates to each other neurons in the adjacent forward layer. In the case that if some of the connection links are missing, then we say that the network is partially connected. BPNN is one of the most important neural networks that can be used as composition controller of a reactive distillation process. The training of BPNN we use is the steepest gradient descent method. The learning factor and convergence of BPNN depend on various parameters such as initial weights, learning rate, momentum factor, and number of training data and hidden layer neurons [1–12].

5.3.1 Initial weights

Initial weight of the neural network is one of the most important parameters, and they are initialized by some small random values that lie in certain defined interval. The initial values of these weights were selected randomly and they affect how slow or fast network will converge. The initial weights cannot be very high due to the use of sigmoidal transfer functions which cause the saturation, and it may be the possibility to stuck at local minima.

5.3.2 Learning rate

Learning rate affects the convergence of the network, small value of learning rate may slow the convergence in case of plateau region, and large value of learning rate may cause fast convergence but arises the problem of overshooting.

5.3.3 Momentum factor

One major problem of gradient descent learning is very slow due to small learning rate and oscillates if the learning rate is too large. To avoid these oscillations, use the momentum factor to the normal gradient. A momentum factor is used with incremental as well as batch mode of training.

5.3.4 Training data

A thumb rule about training data is that it should cover the entire input space such that at the time of training, the training data should be selected randomly. If the training dataset are divided into L number of clusters and T is the lower bound on the number of training pattern, the relation $(T/L) \gg 1$ holds. Sometimes normalization and scaling are also used.

5.3.5 Number of hidden-layer neurons

The size of hidden layer is an important parameter in multilayer feedforward network, which can be determined experimentally. The number of hidden neurons can be determined using a parametric study or method of clustering. If we consider the range of neurons in the hidden layer from [2,20], then start from 2 and verify that if the number of neurons n gives minimum error, then this n will solve the purpose [6–10].

5.4 Incremental mode training of BPNN

For the training of BPNN, let us consider the following notations: input and output patterns of the input layer be $I = (I_{I1}, I_{I2}, \ldots, I_{IM})$ and $I = (I_{O1}, I_{O2}, \ldots, I_{OM})$, where I_{Ii}

and I_{Oi} are input and output at the ith unit at the input layer, respectively. Similarly, $H = (H_{I1}, H_{I2}, \ldots, H_{IN})$ and $H = (H_{O1}, H_{O2}, \ldots, H_{ON})$, where H_{Ij} and H_{Oj} are input and output at the jth unit at the hidden layer, respectively, also let $O = (O_{I1}, O_{I2}, \ldots, O_{IK})$ and $O = (O_{O1}, O_{O2}, \ldots, O_{OK})$, where O_{Ik} and O_{Ok} are input and output at the kth unit of output layer, respectively, and targets are denoted as $T = (T_{O1}, T_{O2}, \ldots, T_{OK})$, where $1 \le i \le M$, $1 \le j \le N$, and $1 \le k \le K$. Also denotes v_{ij} is the connecting weight between ith unit of the input layer to the jth unit of the hidden layer, and w_{jk} is the connecting weight between jth hidden unit to the kth output unit as shown in Figure 5.2. Transfer function at the input layer is linear, at hidden layer logistic sigmoid (eq. (5.17)) and hyperbolic tangent function (eq. (5.18)) at output layer of BPNN.

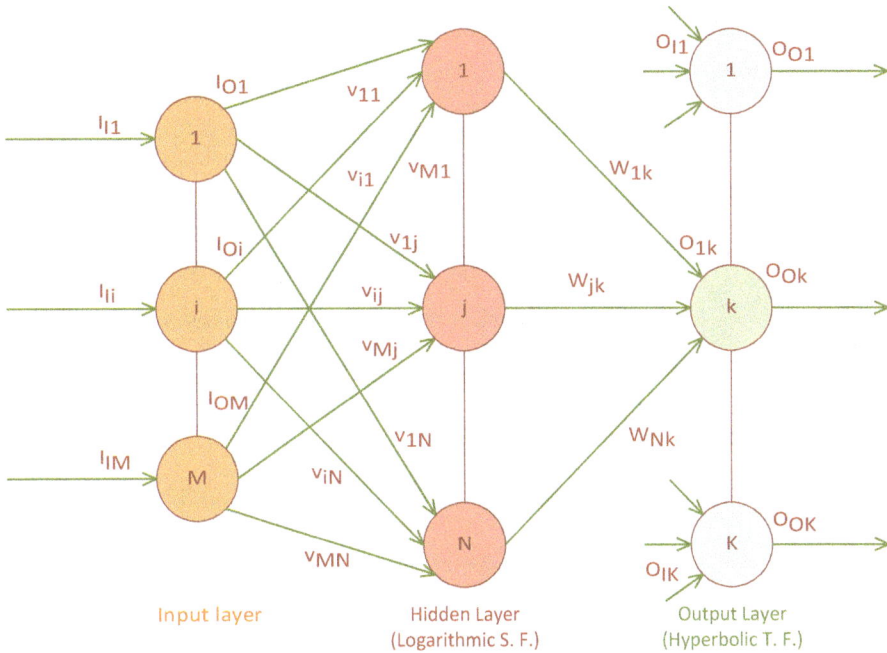

Figure 5.2: Backpropagation neural network with single hidden layer.

The steepest gradient method of learning for BPNN weight adjusting at the hidden and output layers is given by $\Delta v_{ij} = \alpha H_{Ij} f'(x) I_{Oi}$ and $\Delta w_{jk} = \alpha (T_{Ok} - O_{Ok}) g'(x) H_{Oj}$, respectively:

$$f(x) = \frac{1}{1 + e^{-a_1 x}} \tag{5.17}$$

$$g(x) = \frac{e^{a_2 x} - e^{-a_2 x}}{e^{a_2 x} - e^{-a_2 x}} \tag{5.18}$$

Let us define the total error for the P training set (I, T), as in eq. (5.19):

$$E = \frac{1}{P}\sum_{k=1}^{P}(T_{Ok} - O_{Ok})^2$$

$$E = \frac{1}{P}\sum_{k=1}^{P}(E_k)^2, \text{ where } E_k = T_{Ok} - O_{Ok} \tag{5.19}$$

To minimize the cost function, the incremental mode of training is performed in two steps:

Step I: To update the connecting weight between hidden and output layers:

$$w_{jk}(\text{updated}) = w_{jk}(\text{previous}) + \Delta w_{jk} \tag{5.20}$$

where

$$\Delta w_{jk}(t) = -\eta \frac{\partial E_k}{\partial w_{jk}}(t) + \alpha' \Delta w_{jk}(t-1) \tag{5.21}$$

where the first term of RHS of eq. (5.21) is the steepest gradient term and the second term is known as the momentum term. Now evaluate the gradient:

$$\frac{\partial E_k}{\partial w_{jk}} = \frac{\partial E_k}{\partial O_{Ok}}\frac{\partial O_{Ok}}{\partial O_{Ik}}\frac{\partial O_{Ik}}{\partial w_{jk}} \qquad \text{(using chain rule at the } k\text{th neuron)} \tag{5.22}$$

where

$$\frac{\partial E_k}{\partial O_{Ok}} = -(T_{Ok} - O_{Ok}) \tag{5.23}$$

$$\frac{\partial O_{Ok}}{\partial O_{Ik}} = a_2(1 + O_{Ok})(1 - O_{Ok}) \tag{5.24}$$

$$\frac{\partial O_{Ik}}{\partial w_{jk}} = H_{Oj} \tag{5.25}$$

Using eqs. (5.23)–(5.25), in eq. (5.22), we get

$$\frac{\partial E_k}{\partial w_{jk}} = -a_2(T_{Ok} - O_{Ok})(1 + O_{Ok})(1 - O_{Ok})H_{Oj} \tag{5.26}$$

$$\frac{\partial E}{\partial w_{jk}} = \frac{1}{P}\sum_{k=1}^{P}\frac{\partial}{\partial w_{jk}}(T_{Ok} - O_{Ok})^2$$

Step II: To update the connecting weight between input and hidden layers:

$$v_{ij}(\text{updated}) = v_{ij}(\text{previous}) + \Delta v_{ij} \tag{5.27}$$

where $\Delta v_{ij} = -\eta \left\{ \frac{\partial E}{\partial v_{ij}} \right\}_{avg}$ and $\left\{ \frac{\partial E}{\partial v_{ij}} \right\}_{avg} = \frac{1}{P} \sum_{k=1}^{P} \frac{\partial E_k}{\partial v_{ij}}$

$$\frac{\partial E_k}{\partial v_{ij}} = \frac{\partial E_k}{\partial O_{Ok}} \frac{\partial O_{Ok}}{\partial O_{Ik}} \frac{\partial O_{Ik}}{\partial H_{Oj}} \frac{\partial H_{Oj}}{\partial H_{Ij}} \frac{\partial H_{Ij}}{\partial v_{ij}} \text{ (using chain rule at the } k \text{th neuron)} \qquad (5.28)$$

where

$$\frac{\partial E_k}{\partial O_{Ok}} = -(T_{Ok} - O_{Ok}) \qquad (5.29)$$

$$\frac{\partial O_{Ok}}{\partial O_{Ik}} = a_2(1 + O_{Ok})(1 - O_{Ok}) \qquad (5.30)$$

$$\frac{\partial O_{Ik}}{\partial H_{Oj}} = w_{jk} \qquad (5.31)$$

$$\frac{\partial H_{Oj}}{\partial H_{Ij}} = a_1 H_{Oj}(1 - H_{Oj}) \qquad (5.32)$$

$$\frac{\partial H_{Ij}}{\partial v_{ij}} = I_{Ij} \qquad (5.33)$$

Using eqs. (5.29)–(5.33), in eq. (5.28), we have

$$\frac{\partial E_k}{\partial v_{jk}} = -a_1 a_2 (T_{Ok} - O_{Ok})(1 + O_{Ok})(1 - O_{Ok})(1 - H_{Oj}) w_{jk} H_{Oj} I_{Ij} \qquad (5.34)$$

From eqs. (5.26) and (5.34), weight from input to hidden and hidden to output layer is updated [6–12].

5.5 Pseudocode of BPNN

Pseudocode of BPNN comprises the following steps:
1. Randomly initialize weights with entries from $(0,1)$ and learning rate $\eta \in (0.1, 0.9)$
2. While stopping criteria is false
3. For each inputs I_{Ii} and output of input unit I_{Oi} $(1 \le i \le M)$ send it to the hidden unit
4. For each neuron, evaluate $H_{Ij} = \sum_{i=1}^{M} v_{ij} I_{Oi};(1jN)$ and output of hidden unit $H_{Oj} = f(H_{Ij})$
5. For each output neuron $O_{Ok};(1kK)$ evaluated as $O_{Ok} = g(O_{Ik})$; where $O_{Ij} = \sum_{j=1}^{N} w_{jk} H_{Oj}$
6. Error at the kth output neuron is evaluated as $E_k = (T_{Ok} - O_{Ok}) g'(O_{Ik})$ and error at the jth hidden neuron $\delta_j = \sum_{k=1}^{K} w_{jk} E_k f'(H_{Ij})$
7. Calculate the weight change at hidden and output layers as eqs. (5.21) and (5.27)

8. Weight update at the hidden and output neurons is given by

$$v_{ij}(\text{updated}) = v_{ij}(\text{previous}) + \Delta v_{ij} \text{ and}$$

$$w_{jk}(\text{updated}) = w_{jk}(\text{previous}) + \Delta w_{jk}$$

9. Check the stopping criteria, which is either maximum iteration or mean square error (MSE) is less than tolerance

5.6 Teaching–learning-based optimization

Teaching–learning-based optimization (TLBO) is a gradient-free and stochastic population search-based algorithm developed by R. V. Rao in 2011. Inspiration of TLBO is knowledge transfer in a classroom environment; it is a metaheuristic method that works on the effect of teacher on the learners. If we look at the classroom environment, teaching happens in two phases: (1) teacher phase and (2) learner phases.

In teacher phase, students learn from the teacher and in learner phase students learn among themselves and try to increase their knowledge. Teacher phase algorithm uses the best and mean solution of the class, which is from randomly generated population. After this, apply greedy selection, that is, if the new solution generated so far is better than the previous one, then this will be added in the population, and worst solution will be removed from the population. The learner phase has a different variation operator to generate a new solution; for this, select a random population called as partner solution and again apply the greedy method. This is the basic terminology behind the TLBO, and the pseudocode of this method is given by the algorithm.

5.6.1 TLBO algorithm

1. Define optimization function $f(X)$, number of iterations (N_p), lower and upper bound (lb and ub), and maximum iteration T
2. Initialize random population of size N_p
3. For each element of population P, evaluate the value of objective function
   %% **Teacher phase**
4. For $i = 1:T$
   For $j = 1:N_p$
   Using greedy selection choose X_{best}
   Evaluate mean vector from population, i.e., X_{mean}
   $X_{\text{new}} = X_j + r(X_{\text{best}} - T_f X_{\text{mean}})$, where r is random number and $T_f = 1$
   Apply corner bound on and evaluate the objective function value at X_{new}
   If $f(X_{\text{new}})$ is better than $f(X_j)$, then accept X_{new}; otherwise, reject it
   %% **Learner phase** [8–12]

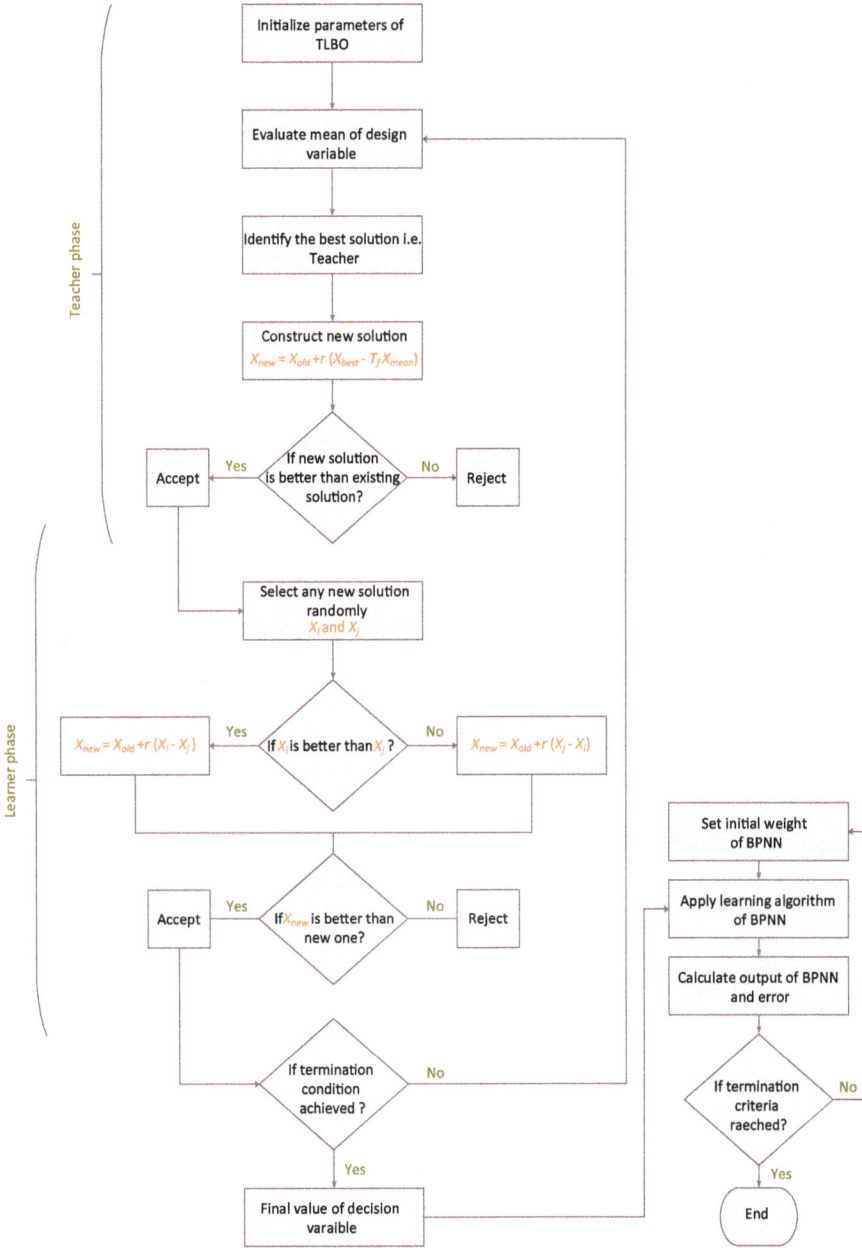

Figure 5.3: Flowchart of TLBO-NN.

5. Select any random solution from population P (say)X_p
6. Again evaluate X_{new} as
   If $f(X_j) < f(X_{new})$

$$X^j_{new} = X_j + r(X_j - X_p)$$

   Else

$$X^j_{new} = X_j - r(X_j - X_p)$$

   If end
7. Apply corner bound on X_{new} and evaluate $f(X_{new})$
8. Accept X_{new} if it is better than X_j
9. End (first For)
10. End (second For)

5.7 Teaching–learning-based optimized neural network (TLBO-NN)

A new learning of neural network has been adopted using TLBO. The MSE is used as fitness function for minimization using TLBO. Once MSE is minimized to certain degree of accuracy then we assign the solution vector to weight of neural network as shown in Figure 5.3, and the weight is updated using TLBO approach. The procedure adapted in TLBO-neural network is presented below:

a. Use TLBO algorithm to assign the initial weight of neural network
b. Apply learning algorithm for weight updation
c. Calculate error using neural network output and target
d. Check the stopping criteria
e. If stopping criteria satisfied, then stop
f. Otherwise, GOTO step 2 [8–12].

References

[1] R.V. Rao, V.J. Savsani, D.P. Vakharia, Teaching–learning-based optimization: A novel method for constrained mechanical design optimization problems, Computer-Aided Design, 43, 2011, 303–315.
[2] U. Chakraborty, S. Roy, Soft Computing, Publisher: 2013, January Pearson, India, ISBN: 9789332524224.
[3] S.N. Sivanandam, S.N. Deepa, Principles of Soft Computing, 2nd edition, Publisher: John Wiley & Sons, ISBN: 9788126527410.
[4] A. Bonarini, F. Masulli, G. Pasi, Soft Computing Applications, 2011, Springer.
[5] B. Widrow, M.E. Hoff, "Adaptive switching circuits", In Proceedings WESCON, pp. 96–104, 1960.

[6] S. Das, A. Biswas, S. Dasgupta, A. Abraham, Bacterial Foraging Optimization Algorithm: Theoretical Foundations, Analysis, and Applications, Foundations of Computational Intelligence Vol. 3, Studies in Computational Intelligence, Vol. 203, 2009, Springer, Berlin, Heidelberg.

[7] S.S. Chouhan, A. Kaul, U.P. Singh, Bacterial foraging optimization based radial basis function neural network (BRBFNN) for identification and classification of plant leaf diseases: an automatic approach towards plant pathology, IEEE Access.

[8] S.N. Sivanandam, S.N. Deepa, Principle of Soft Computing, 2nd edition, 2011, Wiley, India.

[9] S.N. Sivanandam, S.N. Deepa, Neural Network, 2nd edition, 2011, Wiley, India.

[10] V. Sakhre, S. Jain, V.S. Sapkal, D.P. Agarwal, Fuzzy counter propagation neural network control for a class of non-linear dynamical systems, Computational Intelligence and Neuroscience, 2015, 2015, 1–12.

[11] K.S. Narendra, K. Parthasarathy, Neural networks and dynamical systems, International Journal of Approximate Reasoning, 6, 1992, 109–131.

[12] A. Kroll, H. Schulte, Benchmark problems for nonlinear system identification and control using Soft Computing methods: Need and overview, Applied Soft Computing Journal, 25, 2014, 496–513.

Chapter 6
Advanced control of reactive distillation column

6.1 Soft sensors for reactive distillation process

Online measurement of product composition is very important for successful control of reactive distillation process. The measurement using online composition analyzers are difficult because they are expensive, require more maintenance and frequent calibration, and usually accompanied with undesirable time delays. This unusual time delay leads to delayed control action and affects the quality of the product. Therefore, online composition analyzers do not prove to be an effective solution for the accurate measurements. These difficulties can be overcome by using inferential soft sensing techniques. These inferential measurement techniques have recently gained momentum as viable alternatives to hardware sensors. Soft sensors can easily and economically adapt the input–output behavior of the process by minimizing the mean square error (MSE) between the network output and the target output. These sensors are trained in such a way that these can predict the output of the system within a fraction of time. The soft sensors are dynamic models and can be updated at any time. Neural network-based soft sensor which can be modified using steepest descent method to minimize MSEs and fuzzy competitive learning-based counterpropagation soft sensors are proven for reactive distillation process. These soft sensors can be used in the inferential control of reactive distillation process. The performance of the inferential controllers as soft sensors can be analyzed based on the appropriate input selection and other variables.

The modified neural network-based soft sensor and fuzzy competitive learning-based counterpropagation soft sensor estimate the composition of a product using secondary process variable measurements. As the neural network acquires the information with the help of supervised learning, this soft sensor requires a set of secondary measurements and the respective product composition data of the reactive distillation process for training. The input–output data of the reactive distillation process which can be the real-time experimental synthesis and primary simulation process model can be used for identifying the real-time various input conditions. These inputs can be used for training or learning of the soft sensors.

The other suitable soft sensors are adaptive linear neuron, backpropagation neural network (BPNN) and teaching–learning-based optimized neural network (TLBONN) for such nonlinear dynamical control. BPNN based on backpropagation learning are always proven to be a successful soft sensor for many nonlinear dynamical controls. TLBONN is a metaheuristic method based on a stochastic population search which is also proven to be an effective soft sensor for most of the nonlinear processes [1–10].

https://doi.org/10.1515/9783110656268-006

6.2 Inferential control of reactive distillation

Soft sensor for online composition measurement plays an important role in the successful control of the reactive distillation process. Soft sensing techniques can be efficiently used as an alternative to the composition analyzers. Tray temperatures of the column can be chosen as the secondary process variable for the estimation of product composition of the reactive distillation process. The temperature profiles and the corresponding product compositions can be obtained by the sensitivity analysis and simulation of the mathematical model of the reactive distillation process. These temperature profiles used as input to the neural network and composition profiles are output from the neural network [11–14].

6.2.1 Single-loop control strategy

The block diagram of the single-loop control strategy is shown in Figure 6.1. As shown in the figure, packed reactive distillation column has three feed inputs, namely, two feed flow rates and one reflux flow. These feed inputs are feed variables and are disturbing quantities. Sensitive tray temperature can be selected as a secondary variable to estimate the product composition. The selection of trays is a crucial task. Most of the reactions which can be performed in reactive distillation are exothermic in nature; hence, the trays covering the reactive section will be more temperature sensitive. Apart from this, the bottom tray which is connected to the reboiler is also sensitive to temperature variation. Also, the top tray will be enriched in vapor content and hence this is also giving maximum temperature variation in the column.

The output temperature of the sensitive trays can be the input to the selected neural network. Once the number of inputs is passed to the neural network-based control loop, the neural network will then give actual output based on the training algorithm. The sensed output will be compared with the reference composition, and the error obtained will be passed to the controller which designed to compensate the error and will give the corrected output by adjusting the manipulated variable. The disturbance rejection and set point tracking are performed by varying the various feed input to the column.

6.2.2 Cascade control strategy

The block diagram of the cascade control strategy for reactive distillation column is shown in Figure 6.3. The single-loop control strategy uses only one control loop, while in case of cascade control loop strategy there are two loops, mainly for temperature control and composition control.

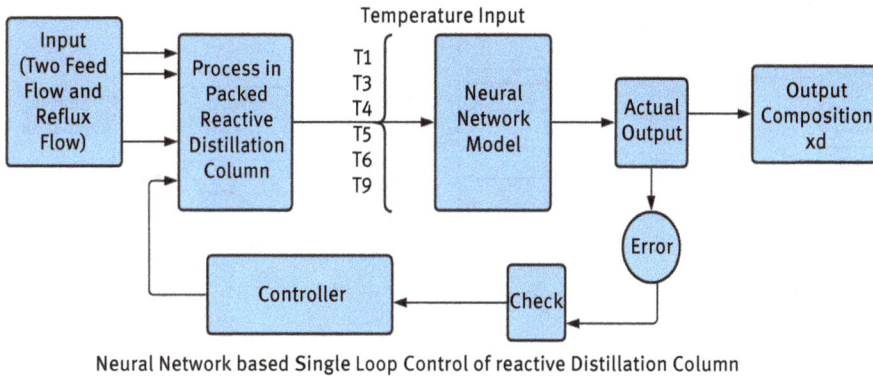

Neural Network based Single Loop Control of reactive Distillation Column

Figure 6.1: Neural network-based single-loop control of reactive distillation [53].

The cascade control loop also uses four inputs as feed rate of reactant A and feed rate of reactant B, the reflux rate, and reboiler heat duty to the reactive distillation column as shown in figure 6.2. The temperature data of the sensitive plates as output of the reactive distillation will be the input to the neural network model. The reboiler heat duty and the temperature of the bottom plate can be the manipulating variables to operate both the cascade loops. The actual output will be compared with the reference value and this will be used to maintain the bottom plate temperature. The reference temperature of the bottom plate will then be compared with the actual temperature of the bottom plate inside the column. This cascade control scheme is planned to measure the most sensitive bottom plate temperature, and the error will be corrected to give the correct composition using the outer loop. And to achieve this, the inner loop must be faster than the outer loop. As compared to single loop, the cascade control scheme is more efficient in producing effective dynamic behavior as it uses two controllers to measure the error generated in both the loops [49–54].

6.2.3 Disturbance tracking

The control strategy should be capable of handling real-time disturbances caused by process change and other factors such as environmental disturbances. The neural network-based control strategy for reactive distillation processes can be tested for any changes in feed flow rates and other input parameters. The changes can be tested for variation of ±10% to check the output behavior with time. The trained neural network is capable of handling nonlinear disturbances and performs well for all such dynamic behaviors. The cascade control strategy can perform better as compared to the single-loop control strategy. The reason for this is there are two loops performing together to make the necessary corrections in the error generated and will be able to track the disturbances more efficiently and quickly. However,

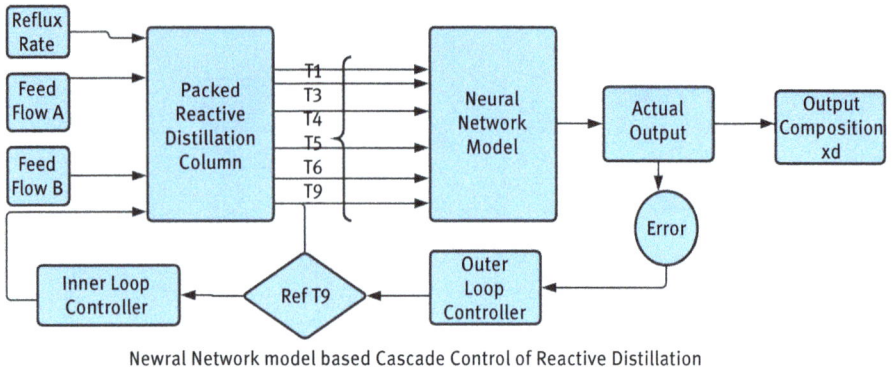

Newral Network model based Cascade Control of Reactive Distillation

Figure 6.2: Neural network-based cascade control of reactive distillation [53].

the single loop is also good for all the nonlinear chemical processes, especially reactive distillation process [53].

6.2.4 Case study

Dual reactive distillation column is the best example of studying single-loop and cascade loop control strategies. The input to the reactive distillation column is two feeds, mainly oil and ethanol to give biodiesel as the final product. The flow rate of each component along with the reflux flow rate and reboiler heat duty is selected to observe the column temperature. Six temperatures, mainly top temperature T1, reactive section's temperatures T3, T4, T5, and T6, and reboiler temperature T9 are recorded and sent as input to the neural network. Based on the six temperatures, the desired product composition is obtained as output from the neural network. The composition variation with time is shown in Figure 6.3. As shown, we can observe that the single-loop control strategies take around 2,500 s to reach constant composition, while double-loop strategy gives constant composition throughout in no time [44–53].

6.2.5 Disturbance rejection and set point tracking

The performance of both single-loop and cascaded loop control strategies were tested by applying various disturbances in the system. The ±10% variation in reflux ratio is introduced as input to the system. The estimations obtained for single-loop and cascade control for positive and negative disturbances are shown in Figures 6.4 and 6.5, respectively. It is observed that in case of single-loop control strategy, the variation in composition is large, that is, the set point changed from 0.96 to 0.9605, and after 600 s, it comes to steady state (set point). In case of cascaded control, the

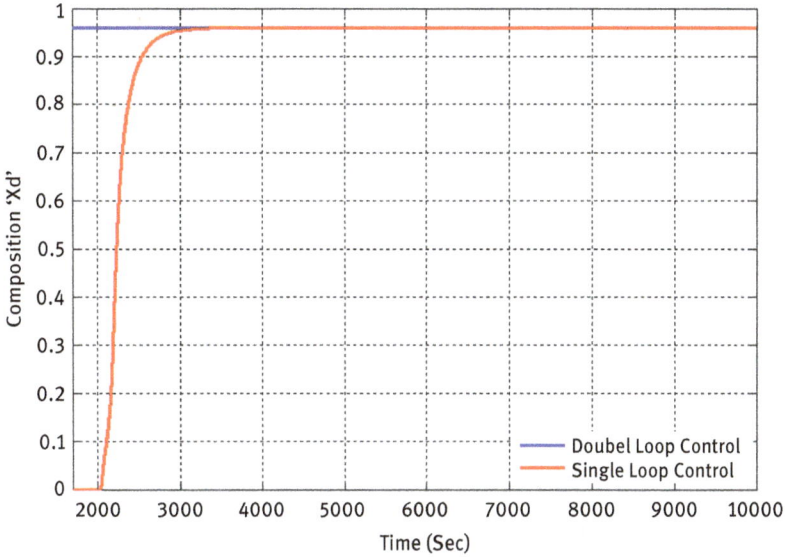

Figure 6.3: Composition variation against time [53].

variation is very small and the set point changed from 0.96 to 0.9605, and only after 50 s, it reaches steady state.

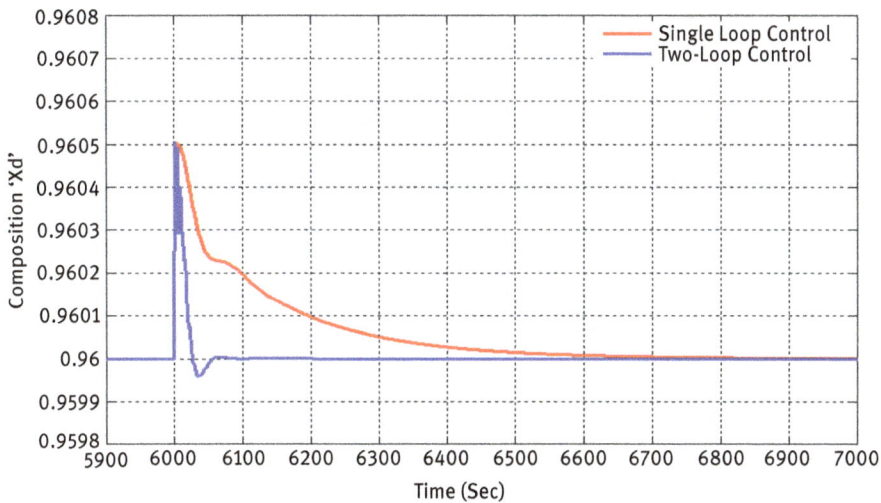

Figure 6.4: +10% variation in reflux ratio [53].

Similarly, when a step change of −10% in reflux ratio is given, in case of a single-loop strategy, again the variation is large, the set point changed from 0.96 to 0.9595, and

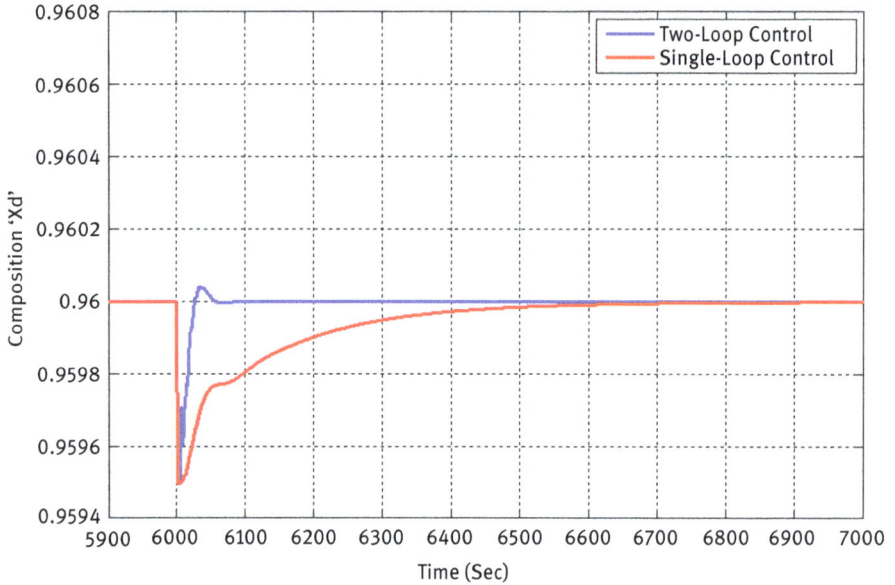

Figure 6.5: −10% variation in reflux ratio [53].

then after 500 s, it came back to steady state. In case of cascaded control, the set point changed from 0.96 to 0.9596, and just after few seconds it reaches steady state.

In the same manner, variation of ±20% in feed flow rate of component A (FA) is observed. The variation is shown in Figures 6.6 and 6.7, respectively. As shown in Figure 6.6, when step change of +20% in F_A is given, the set point changed from 0.96 to 0.983, and after 1,600 s, it reached to steady state while in case of cascaded strategy, the set point changed from 0.96 to 0.964 and then after 800 s only, it reached to steady state without many fluctuations. Similarly, when the set point change of −20% is given in the input feed flow rate FA, the set point changed from 0.96 to 0.94, and after 1,600 s, it reached to steady state, while in case of cascaded control, the set point changed from 0.96 to 0.964 and just after few seconds it reached to steady state.

A similar response is observed when positive variation in set point of feed flow rate F_B changed from 0.035 to 0.042 (+20%), and negative variation in the set point of F_B from 0.035 to 0.028 (−20%) is given and shown in Figures 6.8 and 6.9, respectively. Again, it is clear from the figure that the dynamic response of cascaded control strategy is excellent as compared to the single loop. This is also validated when mixed variation in both feed flow rates F_A and F_B is given, and a response for cascaded control obtained is perfect as compared to the single loop [53] [16–18]. This variation is shown in Figure 6.10.

The tuning of the controller is performed using Tyreus–Luyben tuning method for the various set point changes given to the system, and the performance of the

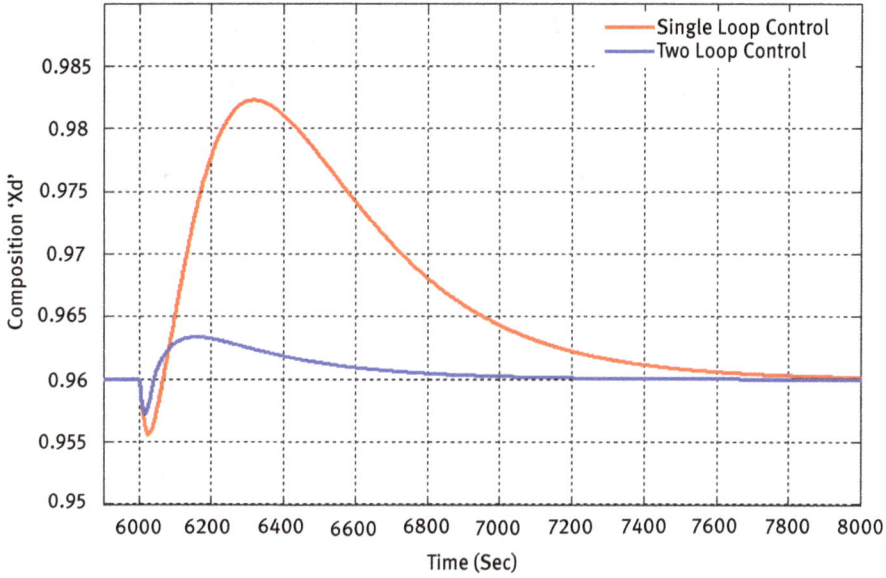

Figure 6.6: +20% (0.035–0.042 L/min) variation in FA [53].

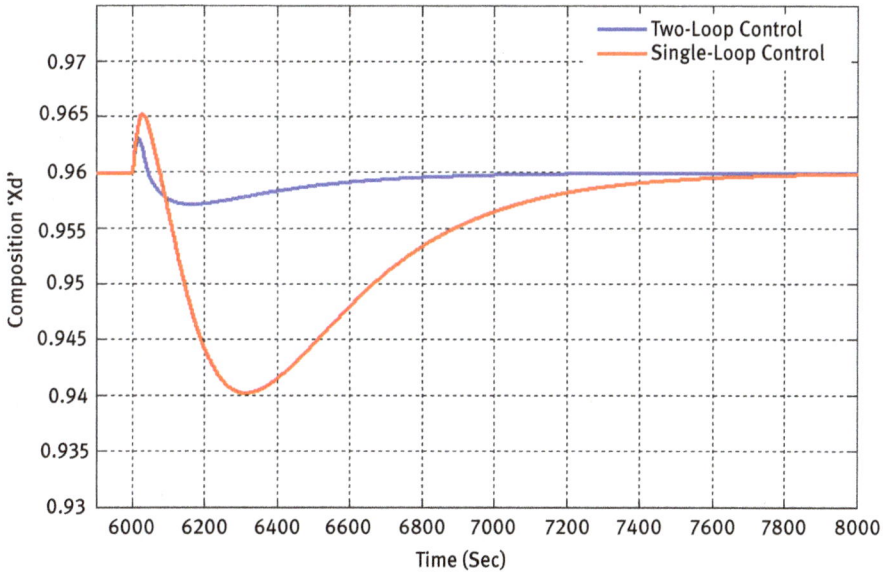

Figure 6.7: −20% (0.035–0.028 L/min) variation in FA [53].

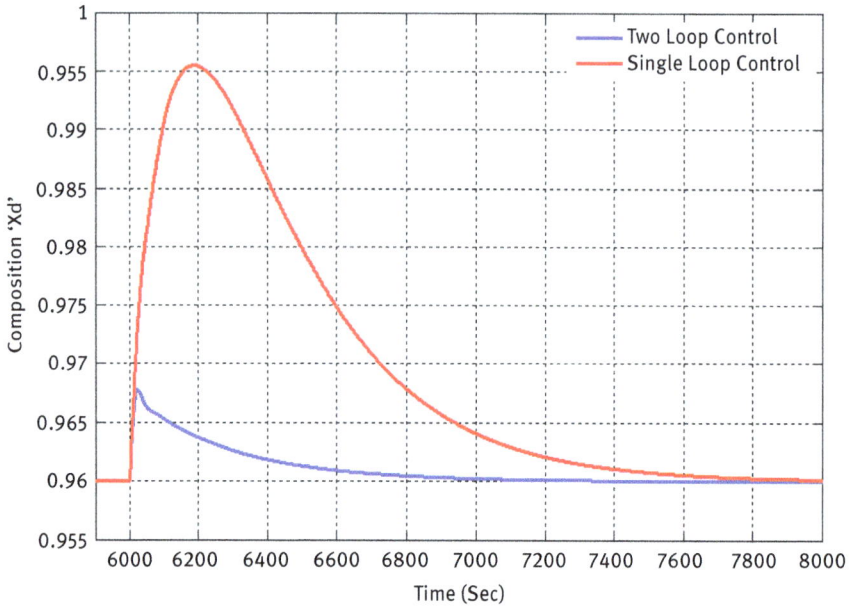

Figure 6.8: +20% (0.035–0.042 L/min) variation in F_B [53].

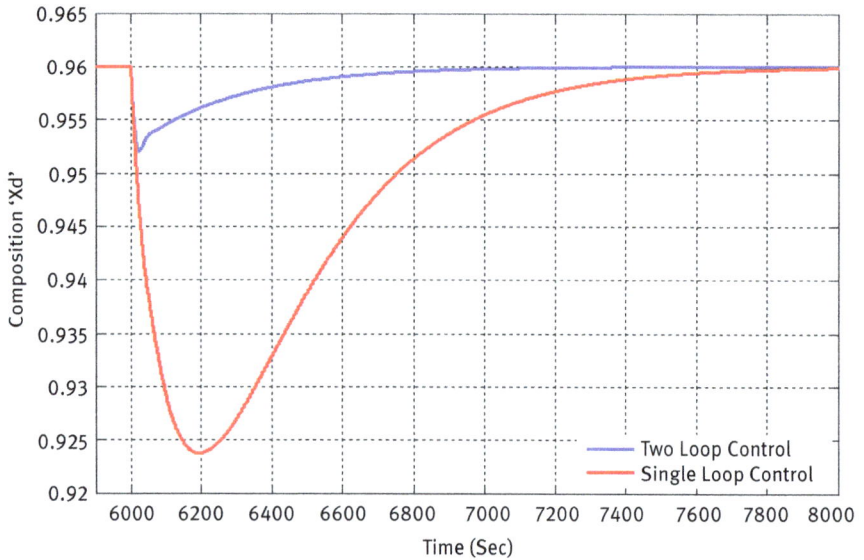

Figure 6.9: −20% (0.035–0.028 L/min) variation in F_B [53].

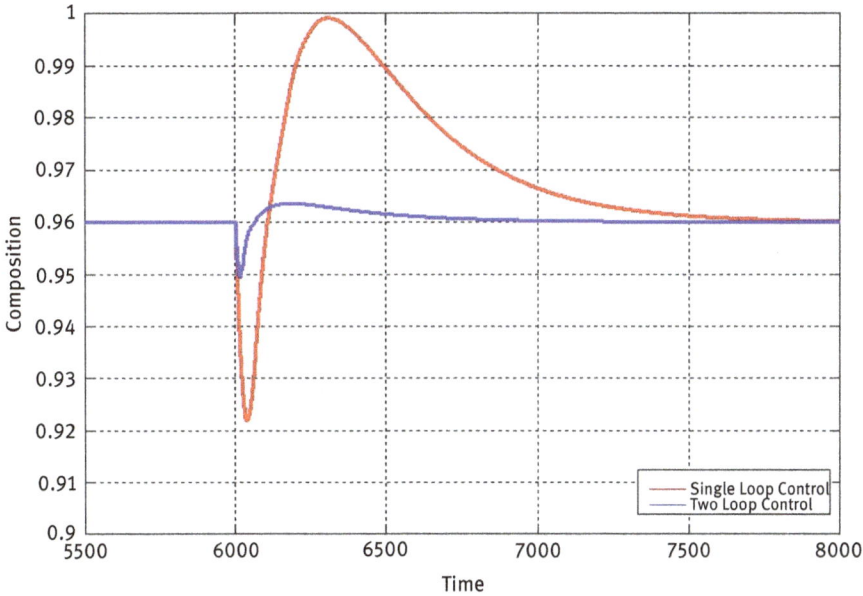

Figure 6.10: Composition variation against mixed variation in flow (F_A + F_B) [53].

controller was evaluated using Integral Absolute Error (IAE) error obtained [16–18]. The IAE error for various set point changes is tabulated in Table 6.1.

Table 6.1: IAE of PID controller subjected to various set point changes.

Manipulating variable	Tuning rule	Step change	gain	Integral time (s)	Derivative time (s)	IAE error
Reboiler heat duty	Tyreus–Luyben	+10%	205	7	0.85	0.0050
		−10%	185	8.2	1.27	0.0010
Flow rate of feed A	Tyreus–Luyben	+10%	193.3	13.20	0.01	0.0135
		−10%	122.3	19.8	0.52	0.0140
Flow rate of feed B	Tyreus–Luyben	+10%	557.1	13.2	0.16	0.0034
		−10%	537.2	13.2	0.45	0.0054
Reflux ratio	Tyreus–Luyben	+10%	196	9.4	0.04	0.0012
		−10%	185	7.2	0	0.0017

6.2.6 Validation of soft sensor

A pilot-scale reactive distillation column is used to validate the soft sensors. The reactive distillation column has 10 stages, 2 feed tanks each of 10 L capacity and a reboiler of 3 L capacity. Two feed streams are passed from top of the reactive zone (i.e., third stage) and bottom of the reactive zone (i.e., sixth stage), respectively. A total of 10 temperature sensors are used to sense the temperature of each stage, including reboiler. Two rotameters are connected to each feed tank for flow measurement and one rotameter is connected to the condenser for measurement of water flow to the condenser. A reflux divider is used to control reflux ratio and reflux flow.

Experimental temperature profile for different heat input is collected using data logger and is used to test the soft sensors. The stage-wise temperature data obtained for biodiesel synthesis is tabulated in Table 6.2. This 10-temperature data is obtained through 10-temperature sensors and used as input for training and testing the soft sensor. The experimental temperature is plotted with respect to time and presented in Figure 6.11. The output is obtained as composition of the biodiesel and water at the bottom and top of the reactive distillation column, respectively. The experimental composition of biodiesel obtained is 97.3%.

As shown in Figure 6.11, it is observed that the temperature is increasing, and it is maximum at the reboiler. The comparative plot of stage-wise temperature obtained from experimental and simulation is presented in Figure 6.12. As shown in figure, we can observe that there is a good agreement between temperature obtained from experimental and simulation results. The temperature is continuously increasing and the highest temperature is at stage 5, which is identified as the most sensitive stage at which the gain is maximum and then it is decreasing and becomes constant. The comparative graph of composition obtained from experimental and simulation is also plotted and presented in Figure 6.13. The maximum product composition of biodiesel obtained after experimental and simulation is 99%. As shown in figure, there is a good agreement between experimental and simulation composition. The composition of each component on each stage is given in Table 6.3.

The same procedure is followed, and the data obtained for TAEE is tabulated in Table 6.4. The composition of TAEE and water is obtained at the bottom and top of the reactive distillation column, respectively. The temperature profile of TAEE is plotted with time and shown in Figure 6.14. It is clear from the figure that the temperature is continuously increasing and the highest temperature is at reboiler. From the above temperature data, we observe that system exhibits multiple steady states. Temperature firstly increases at the start of the reaction and due to increase in the vapor formation because of continuous supply of heat by providing reboiler duty. The temperature drops after the start of reflux because cool and condensed liquid enters the column. Temperature again increases after the reflux is stopped due to further increase in the reaction extent. This variation can also be studied from the temperature profile of simulation. A continuous increase in the column temperature is observed from top of the

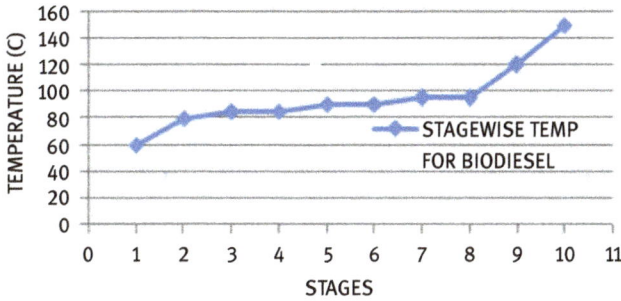

Figure 6.11: Experimental temperature profile for biodiesel synthesis.

Table 6.2: Experimental stage-wise temperature data for biodiesel synthesis.

Tray	1	2	3	4	5	6	7	8	9	10
Time (min)					Temperature (°C)					
10	60	80	83	82.1	86	90.3	98	99	114	137
20	60	81.6	82.7	82.8	86.3	89.9	96	100	113	134
30	61	81	82	82.1	86	90	99	100	114	136.2
40	62	81.4	82.3	82	87.2	90	100	102	117	136
50	62.7	82	82	82.4	87	89.7	100	101	129	140
60	61	82.2	82.5	83.6	87	89.6	101	102	129	140.3
70	61	82	83.4	84	88	90	101	100	120	148
80	60.2	83.9	83.4	84.2	87.8	89.6	101	102	123	150
90	60	82	82	84	88	92	100	102	124	150
100	61	82	82.9	84	87.9	91.5	101	101	121	154
Average	**60.9**	**81.9**	**82.6**	**83.7**	**87.1**	**90.2**	**101**	**102**	**120.7**	**142.6**

column toward the bottom because of exothermic heat release by the reaction. Also, since reboiler is present at the bottom of the column, the highest temperature is recorded as that of the last plate [1–54].

The experimental and simulation composition obtained is plotted and shown in Figure 6.15. It is clear from the figure that there is good agreement between experimental composition and the composition obtained from simulation. The maximum composition obtained in both is 85%. The stage-wise composition is presented in Table 6.5.

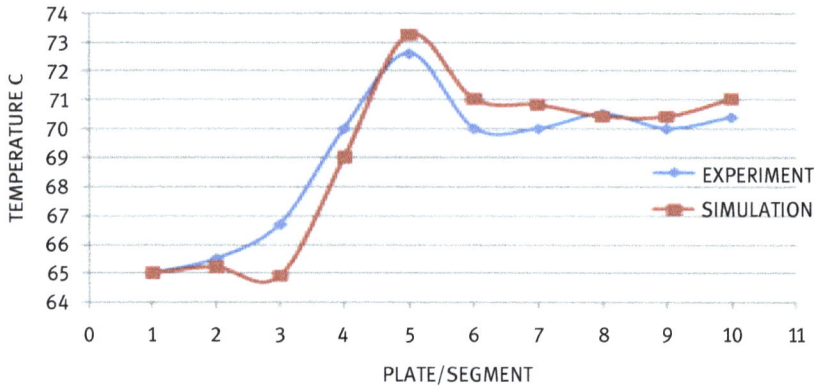

Figure 6.12: Temperature plot obtained from experiment and simulation for biodiesel.

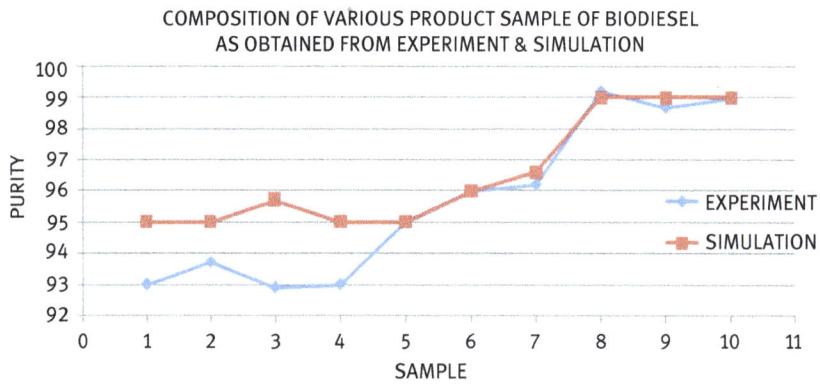

Figure 6.13: Comparative plot of composition obtained from experiment and simulation.

Table 6.3: Product composition of each component on each stage/tray.

Stage/tray	Methanol	Glycerol	Water	Biodiesel
1	0.001	0.2	0.7	0.099
2	0.001	0.12	0.8	0.079
3	0.001	0.08	0.6	0.319
4	0.001	0.1	0.5	0.399
5	0.001	0.149	0.45	0.4
6	0.08	0.24	0.18	0.5
7	0.08	0.10	0.22	0.6
8	0.1	0.10	0.14	0.66

Table 6.3 (continued)

Stage/tray	Methanol	Glycerol	Water	Biodiesel
9	0.2	0.02	0.05	0.73
10	0.005	0.00	0.005	0.99

Table 6.4: Experimental stage-wise temperature data for TAEE synthesis.

Tray	1	2	3	4	5	6	7	8	9	10
Time (min)				Temperature (°C)						
10	82	92	93	96	98	99	99	103	102	102
20	82	92	94	96	96	98	100	103	102	104
30	81	94	94	98	99	98	100	102	103	104
40	85	94	94	99	100	99	102	104	102	103
50	85	98	94	97	100	100	100	102	104	104
60	88	95	95	98	101	101	102	102	104	104
70	87	94	96	98	101	99	102	103	103	104
80	88	95	96	98	100	100	101	104	104	104
90	88	94	96	98	101	101	102	104	104	103
100	88	94	96	98	101	101	102	103	104	104
Average	88	94	96	98	101	101	102	104	104	104

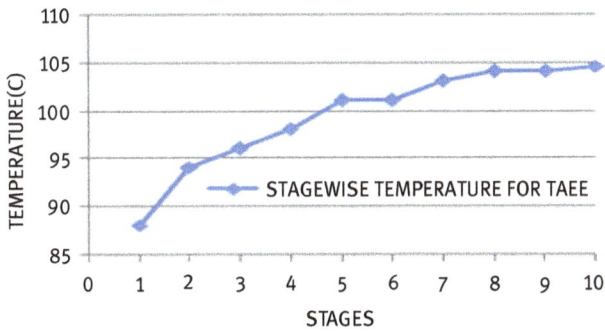

Figure 6.14: Experimental temperature profile for TAEE synthesis.

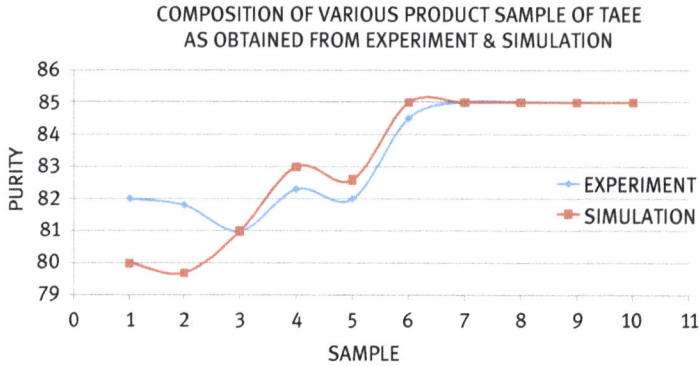

COMPOSITION OF VARIOUS PRODUCT SAMPLE OF TAEE
AS OBTAINED FROM EXPERIMENT & SIMULATION

Figure 6.15: Composition obtained from experiment and simulation for TAEE.

Table 6.5: Product composition of each component on each stage/tray.

Tray	Ethanol	TAA	Water	TAEE
1	0.10	0.10	0.55	0.25
2	0.10	0.14	0.45	0.31
3	0.10	0.20	0.35	0.35
4	0.13	0.17	0.3	0.40
5	0.13	0.11	0.26	0.50
6	0.13	0.13	0.22	0.52
7	0.08	0.16	0.20	0.56
8	0.06	0.17	0.19	0.58
9	0.07	0.1	0.1	0.73
10	0.02	0.065	0.065	0.85

6.2.7 Simulation study of dual RD of biodiesel and TAEE

Steady-state simulation studies can be carried out using the Aspen Plus process simulator. Firstly, the reactive distillation column can be simulated by the module, Radfrac, using nonrandom two-liquid property method. For this purpose, Aspen Plus requires the specification of components, property method, feed conditions (flow rate, composition and thermal state), operating pressure, column configuration (number of stages, feed location, reaction stage, and types of condenser and reboiler), two operating parameters, and reaction type. The two operating parameters can be chosen from a set of parameters such as reflux ratio, distillate rate, bottoms rate, reboiler duty, and condenser duty. The reaction type can be chosen from kinetic, equilibrium, and conversion.

The simulation is performed for biodiesel production using feed flow rate of 0.03 L/min, 50 °C temperature, and 1 atm pressure for ethanol, and 0.05 L/min and 70 °C temperature for oil. The specifications and other results of RDC are included in Table 6.6. The simulation flow sheet is shown in Figure 6.16. The composition profile obtained after simulation for TAEE and biodiesel is shown in Figures 6.17 and 6.18, respectively. As shown in Figure 5.18, the maximum composition of TAEE obtained is 82.6%, while maximum composition of biodiesel obtained is 99% as shown in Figure 6.18.

From the composition profile as obtained from the simulation of reactive divided wall distillation column in Aspen Plus, it is observed that it is possible to obtain the product with 100% purity at bottom as well as side product can also be obtained in a purer form, which is a major advantage of reactive divided wall distillation column. From the simulation result, we can conclude that the reactive divided wall distillation column can emerge as a novel technique for biodiesel synthesis [41–53].

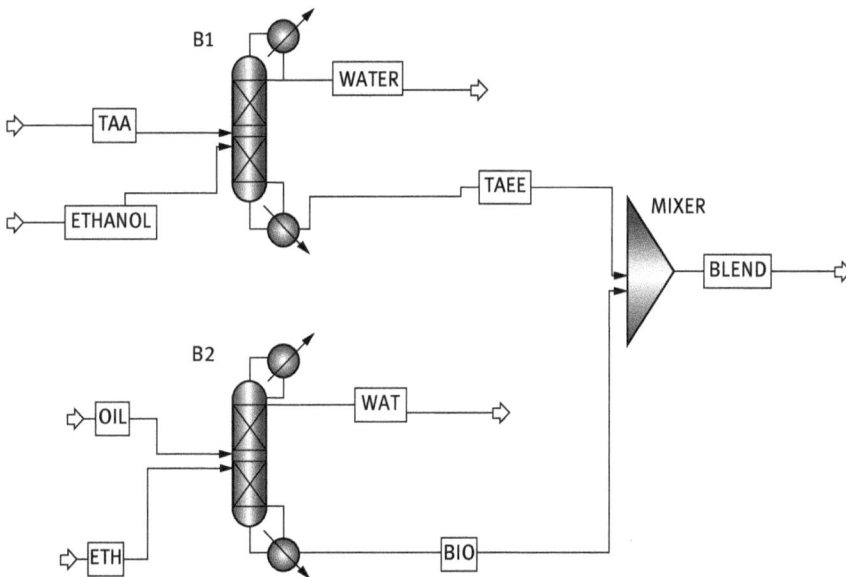

Figure 6.16: Simulation flow sheet of dual reactive distillation column [53].

The sensitivity analysis can be carried out to obtain the optimized value of input design variables corresponding to highest product purity. For biodiesel synthesis process, the boil-up ratio and oil feed flow rate are important parameters for controlling the product composition; hence, these parameters are strictly optimized using a process simulator and the curves for optimized input parameters are shown

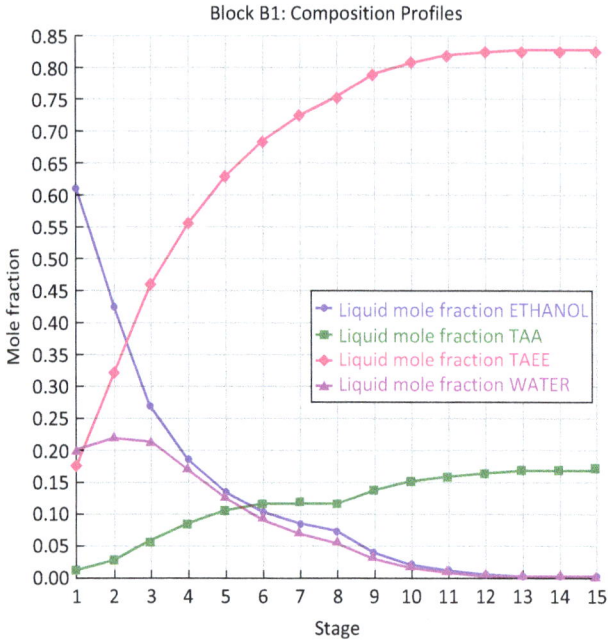

Figure 6.17: Composition profile of TAEE in column RD 1 [53].

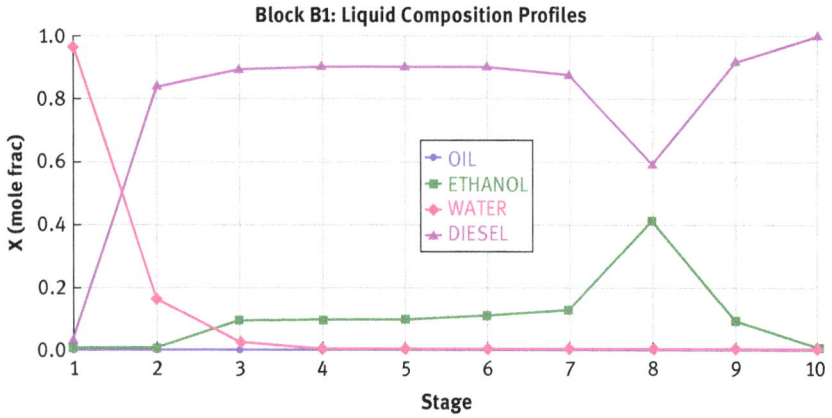

Figure 6.18: Composition profile for biodiesel in RD 2 [53].

in Figures 6.19 and 6.20, respectively. As shown in figure, the product composition is continuously increasing with an increase in the boil-up ratio and it is constant after the boil-up ratio of 0.60. Figure 6.20 presents that approximately 100% product composition is obtained at an oil mass flow rate of 700 kg/h. As oil mass flow rate is increased, the composition of products is decreased. Therefore, the optimized boil-up ratio is 0.65 and optimized oil mass flow rate is 700 kg/h [36–37] [41–53].

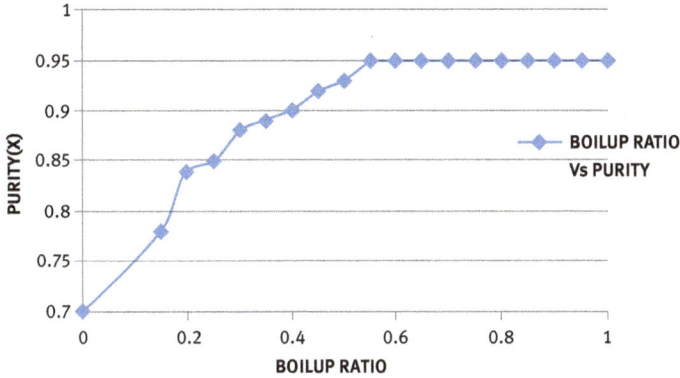

Figure 6.19: Optimization for biodiesel purity versus boil-up ratio.

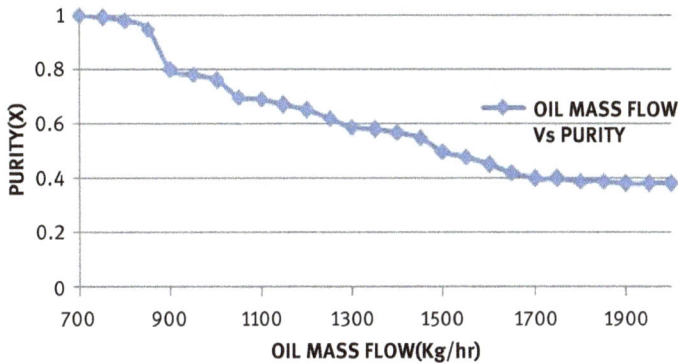

Figure 6.20: Optimization for biodiesel purity versus oil feed flow rate.

Similarly, for TAEE synthesis, the sensitivity analysis is carried out by considering the reboiler duty and reflux ratio as the critical parameters corresponding to highest product composition, and the results are shown in Figures 6.21 and 6.22, respectively. As shown in figure, the product composition is maximum at reboiler duty of 1.2 kW, and further increase in reboiler duty, that is, after 1,400 W, the composition is constantly below 30%. Similarly, at a reflux ratio of 3.2, the product composition is maximized, while a further increase in reflux ratio results in a decrease in

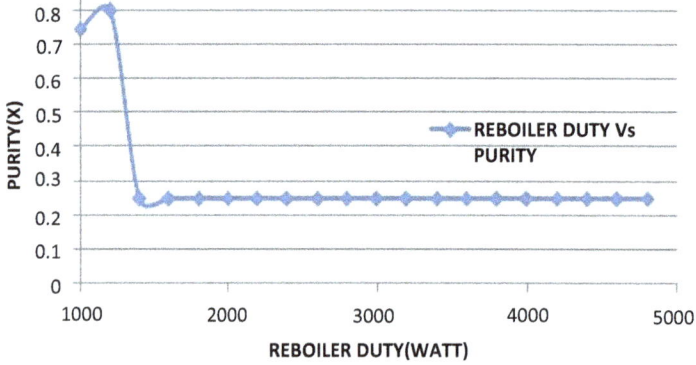

Figure 6.21: Optimization for TAEE purity versus reboiler duty.

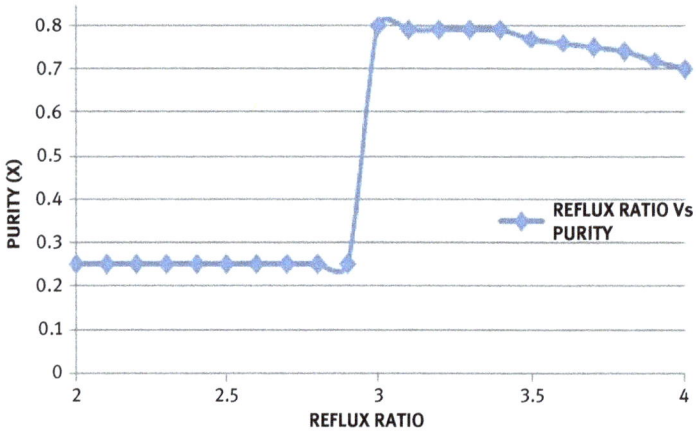

Figure 6.22: Optimization for TAEE purity versus changes in reflux ratio.

composition. Therefore, it is concluded that 1.2 kW is the optimized reboiler heat duty and 3.2 is the optimized reflux ratio [41–53].

6.2.8 Simulation of membrane-assisted RD

Aspen Plus process simulator is used to carry out the simulation of membrane-assisted reactive divided wall distillation column. Two inlets from one side of the reactive divided wall, namely, oil and ethanol, are passed to produce biodiesel using transesterification reaction, while TAA and ethanol as inlet passed through the other side of the

Table 6.6: Operating conditions and results of dual RD integration technique.

Parameters	Column 1 (TAEE)	Column 2 (biodiesel)
Pressure	Atmospheric	Atmospheric
Reboiler duty	2 kW	1.5 kW
Feed 1 temperature	353 K (TAA)	343 K (mustard oil)
Feed 2 temperature	323 K (ethanol)	333 K (methanol)
Feed 1 flow rate	0.02 L/min	0.04 L/min
Feed 2 flow rate	0.02 L/min	0.025 L/min
Reflux ratio	3	
Total stages	8	
Boil-up ratio	0.5	
Rectifying section	1–3	
Reactive zone	3–6	
Stripping section	6–9	
Condenser type	Vertical	
Packing in reactive zone	Amberlyst 15	
Packing in nonreactive zone	Katapak S	
Condenser temperature	330.4 K	
Condenser heat duty	−0.54 kW	
Distillate rate	10.56 mol/h	
Reflux rate	52.81 mol/h	
Reboiler temperature	335.66	
Bottom rate	86.12 mol/h	
Boil-up rate	23.62 mol/h	
Boil-up ratio	0.274	
Activation energy (kcal/mol) for biodiesel		
Forward	13.14	
Backward	6.20	
Specific reaction rate (Lmol/s) for biodiesel		
Forward	3.4×10^8	
Backward	8.3×10^8	

Table 6.6 (continued)

Parameters	Column 1 (TAEE)	Column 2 (biodiesel)
Heat of reaction (kJ/mol) (biodiesel)	100.7	
Heat of vaporization (kJ/kg) (biodiesel)	302.38	
Activation energy (kJ/mol) for TAEE		
Forward activation energy	81	
Backward activation energy	99	
Specific reaction rate (mol/kg s) for TAEE		
Forward reaction rate	0.0064	
Backward reaction rate	0.0015	
Heat of reaction (TAEE) (kJ/mol)	41.708	
Heat of vaporization (TAEE)	31.2 kJ/mol	

divided wall column. The partitioning wall was made up of a membrane which is selective to the common reactant ethanol. Zeolite is used as a membrane which is the most suitable membrane for esterification reactions carried out in a reactive distillation column. After product formation, water as a by-product from both the reactions was collected on top, while biodiesel and TAEE were collected at the bottom. Maximum ethanol, which was recovered through the membrane, was 72%, which after collecting enough can be recycled to reuse as a common reactant for both reactions [41–53]. The composition profile of ethanol recovery obtained after simulation is shown in Figure 6.23.

6.3 Intelligent control of methyl acetate reactive distillation

Methyl acetate was conventionally manufactured in a continuous reactor followed by a separator but with the advent of reactive distillation technology, methyl acetate was the most dominating experiment as well as industrially trailed chemical in this technique. The reason being easy detectability is due to its fruity smell, its vast application in paint, varnish, solvent, and so on, and the last but not the least very optimum range of operating condition required by the system. The main task is again the control of product composition. Control using conventional controllers such as proportional–integral–derivative (PID) controller gives sluggish behavior because reactive distillation comes under highly nonlinear process and this nonlinear behavior is due to the combination of reaction and separation simultaneously. The reaction occurring in case of methyl acetate is esterification reaction and exothermic in nature. Due to this reason, selection of control trays is an important task. Appropriate location of

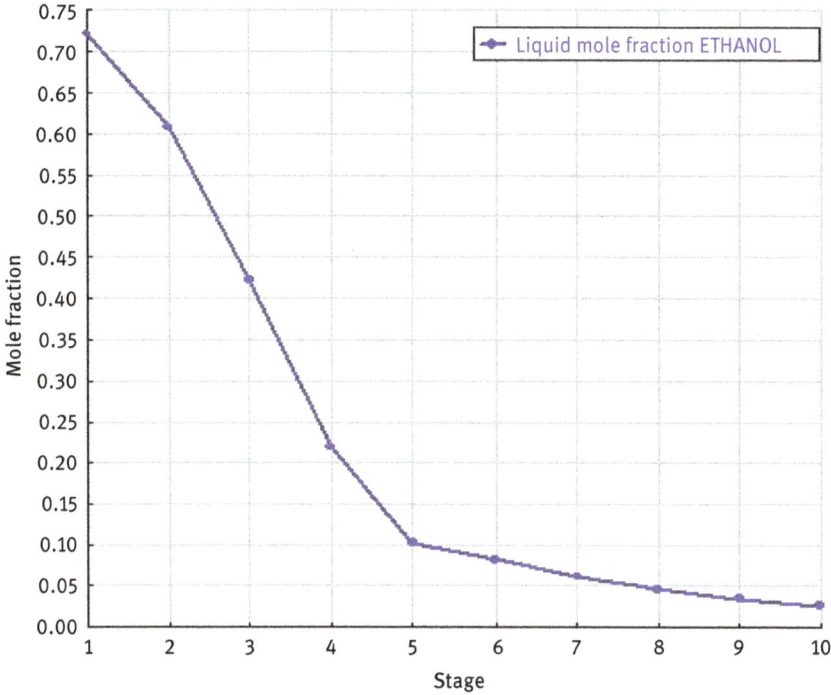

Figure 6.23: Composition profile of ethanol recovery through membrane.

controller is important to have direct control action. In case of methyl acetate reactive distillation, temperature of the selected stages is chosen as a secondary process variable to control the product composition. For this, selection of trays whose temperatures are to be controlled is important for the design of intelligent controller [21–35].

6.3.1 Control structure of methyl acetate reactive distillation

The control structure of methyl acetate reactive distillation column is shown in Figure 6.24. As shown in figure, two feed streams, acetic acid as F_A and methanol as F_B, are entering to the top and bottom of the reactive section. The desired product is collected at the top and water as by-product is collected at the bottom. The process is operated without any excess reactant, and it is essential to manipulate the feed flow rate so that stoichiometry is exactly balanced. The two feed flow rates are manipulated by controlling the respective tray temperature and by this way they are controlling the product composition. For control of two tray temperature, two PID controllers are used. The column pressure is assumed to be constant and maintained by manipulating the condenser heat duty. The level of liquid in the reboiler is controlled by manipulating the bottom product flow rate [21–35].

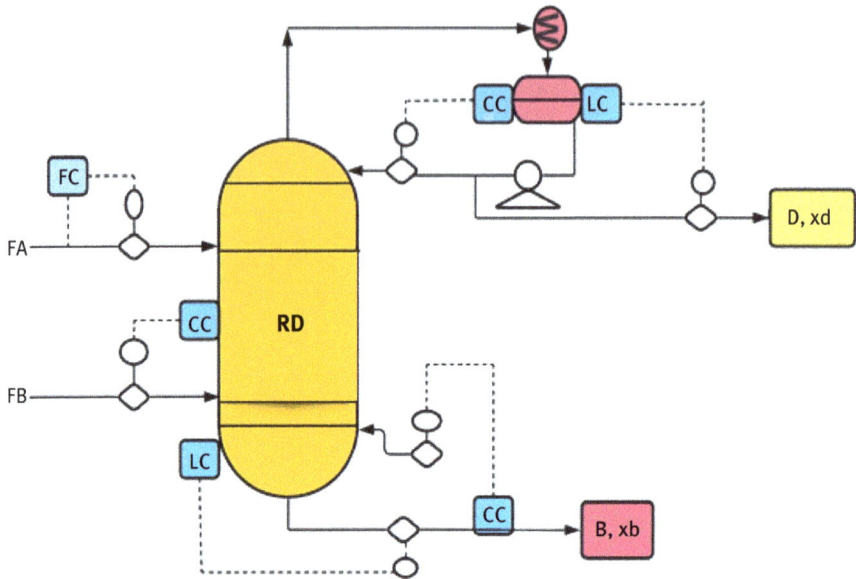

Figure 6.24: Control structure of methyl acetate reactive distillation column.

6.3.2 Identification of control trays

The identification of the control tray is done by gain analysis. A known change says 10% is given to each feed flow rate, and the corresponding change in temperature at each stage is observed. The gain is defined as the ratio of change in temperature to change in flow rate.

The most sensitive tray is identified, which shows the maximum gain. In case of methyl acetate reactive distillation, change in feed flow rate of acetic acid and methanol is given and change in temperature is obtained. Figure 6.25 shows the gain analysis for feed flow rate F_A. It is clear from the figure that stage 3 is identified as the most sensitive tray as it has the maximum gain. Hence, it is important to control the temperature of the tray 3 by manipulating the feed flow rate of component A.

Similarly, the change in feed flow rate of component B is given to identify the sensitive plate for this change. The gain analysis for the same is shown in Figure 6.26. It is clear from the figure that when a change of 10% in feed flow rate of component B is given, the corresponding highest gain is observed at stage 6. Hence, temperature of tray 6 can be controlled by variation in the feed flow rate F_B. As the reaction occurring in the column is exothermic in nature, the temperature profile of the column will be increased due to the addition of heat of reaction. Therefore, to control the product composition, the rate of reaction must be manipulated by controlling the sensitive tray temperature [21–55].

STAGE GAIN CORRESPONDING TO CHANGE
IN FEED FLOW RATE OF ACETIC ACID

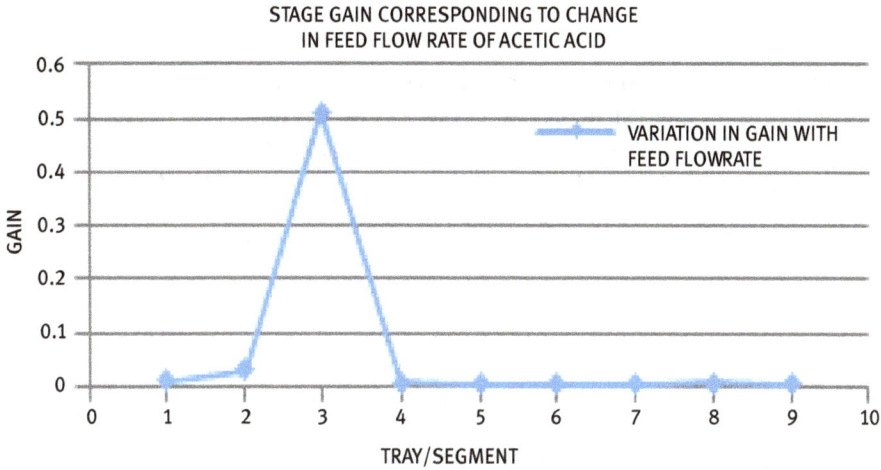

Figure 6.25: Gain analysis for feed flow rate F_A [55].

STAGE GAIN CORRESPONDING TO CHANGE IN
FEED FLOW RATE OF METHANOL

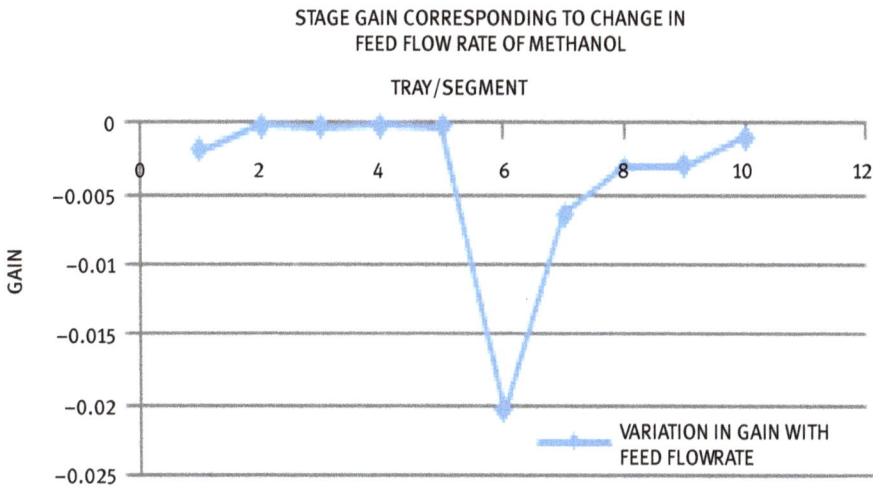

Figure 6.26: Gain analysis for feed flow rate F_B [55].

The product composition is controlled by controlling the tray temperature and regulating the corresponding feed flow rate. The product composition obtained is 96%. To achieve this much conversion, the steady-state simulation is performed to attain the temperature near the input condition. Thus, the desired product composition is achieved by maintaining the temperature to its set point obtained from simulation. Fuzzy counterpropagation neural network (FCPN) is designed for this purpose. For comparison purpose, a conventional PID controller is also designed [21–55].

6.3.3 PID controller

A PID controller is a feedback mechanism-based control and widely used in industries. A PID controller calculates the error value as the difference between a measured process variable and a desired set point. The controller attempts to minimize the error by adjusting the process variable using a manipulated variable. The output of the PID controller can be expressed as follows:

$$u(t) = k_p e(t) + k_i \int\limits_0^t e(\tau)d\tau + k_d \frac{de}{dt}$$

where u is the control signal, e is the error, τ is the integral time, and k_p, k_i, and k_d are proportional, integral, and derivative gain, respectively. All the three control combinations of conventional PID controller are used to control the reactive distillation process [1–4]. The PID controller is tuned using Tyreus–Luyben method. The controller design parameters are tabulated in Table 6.7.

Table 6.7: Controller parameters [55].

Control tray	Gain	Integral time (s)	Derivative time (s)
Third tray	3.29	12.79	0.6
Sixth tray	3.89	10.5	0.5

The designed PID controller is used to control two sensitive plate temperatures by manipulating the respective feed flow rate. The response of PID controller for the third tray and sixth trays is shown in Figures 6.27 and 6.28, respectively. As shown in the figure, it is clear that PID controller initially behaves sluggish and there are oscillations in response as the process is highly nonlinear. To control such a highly nonlinear process, it is desired to use advanced control technique. In this work, we have implemented BPNN, dynamic network (DN), and fuzzy learning-based counterpropagation neural network (FCPN) to control the product composition of reactive distillation process. The performance of these networks is then compared.

6.4 Fuzzy learning-based counterpropagation network (FCPN)

FCPN is designed and implemented for two cases: firstly, it is implemented for nonlinear dynamical system and secondly it is implemented in case of reactive distillation system.

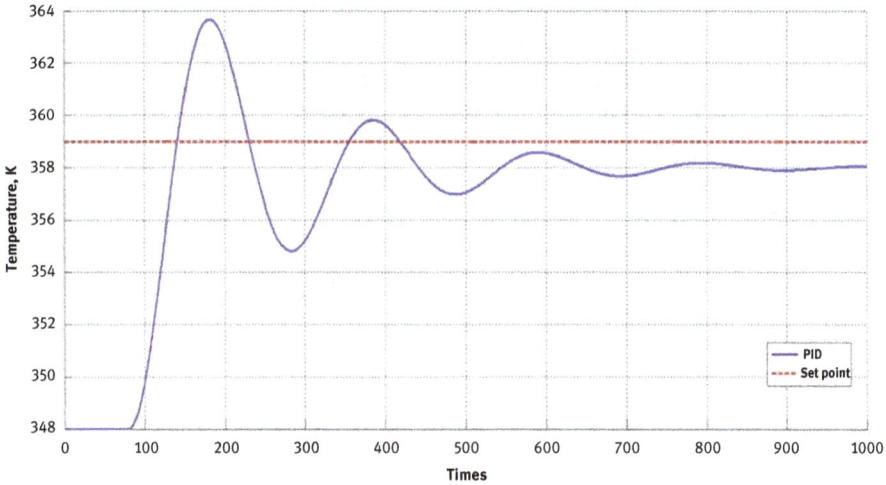

Figure 6.27: Response of PID controller for the third tray temperature [53].

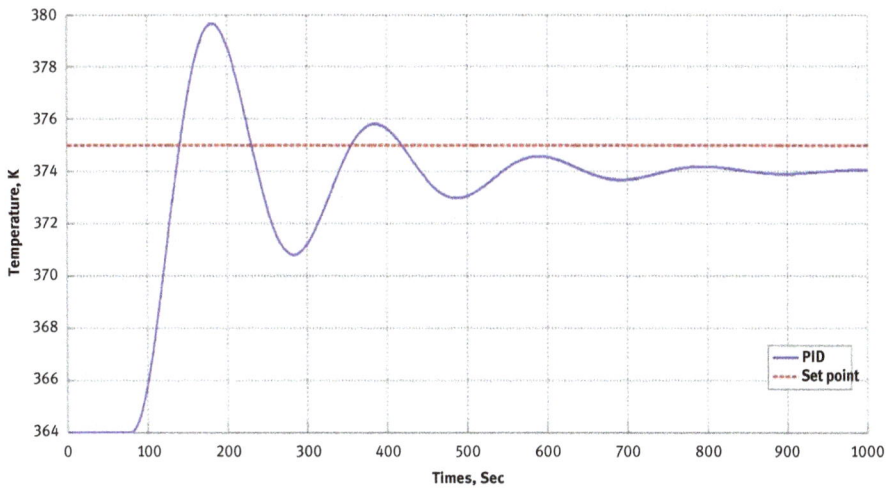

Figure 6.28: Response of PID controller for the sixth tray temperature [53].

6.4.1 FCPN for nonlinear dynamical system

We have designed a FCPN design to control nonlinear dynamical system. In the FCPN, CPN model is trained by FCL algorithms. The FCL learning is used for adjusting weights and updating best matched node in discrete-time nonlinear dynamical system. For this, the approximation for a class of nonlinear dynamical system using FCPN is carried out. The performance criteria of proposed FCPN for nonlinear dynamical system are effectively improved by compensating a reference signal and a controller signal. FCPN is employed to optimize the mean absolute error, MSE, best fit rate (BFR), and so on.

Let us consider the nonlinear model of discrete-time nonlinear dynamical systems for multiple-input–multiple-output. Let $f : R^n \rightarrow R$ and $g : R^m \rightarrow R$ be the nonlinear continuous differentiable functions of the model which are approximated by FCPN to the desired degree of accuracy. Once the system has been parameterized, the performance evaluation is carried out by FCPN. The nonlinear dynamical system is described by the following difference equation and a Box–Jenkins time series data:

$$y(k+1) = f[y(k), \ldots, y(k-n+1); u(k), \ldots, u(k-m+1)]$$

where $[u(k), y(k)]$ represents the input–output pair of the system at time k, and their order is represented by (n, m). FCL is used to learn the system defined for the above equation, and the performance of the FCPN can be measured by the error function given in the following equation:

$$E(k) = \frac{1}{2} [y(k) - \bar{y}(k)]^2$$

where $u(k)$ is the input, $y(k)$ is the system output, and $\bar{y}(k)$ is the neural network output.

The above error function can be written for neural controller as follows:

$$E^c = \frac{1}{P} \sum_{p=1}^{P} [y(k) - \bar{y}^c(k)]^2$$

where $\bar{y}^c(k)$ neural controller output and the above equation are known as minimized error function, where P is the total number of input pattern.

The program is written and simulated in MATLAB and run to demonstrate the effectiveness of the controller for the nonlinear dynamical system. Generalized form of the model of the nonlinear dynamical system can be written as follows:

$$y_p(k+1) = f[y_p(k), y_p(k-1), y_p(k-2), u(k), u(k-1)]$$

where the unknown function f has the following form:

$$f[x_1, x_2, x_3, x_4, x_5] = \frac{x_1 x_2 x_3 x_5 (x_3 - 1) + x_4}{1 + x_3^2 + x_2^2}$$

In the identification model, it is approximated by FCPN, and Figure 6.29 shows the output of the system. When the identification procedure of the model is carried out, the random input signal is uniformly distributed in the interval $[-1, 1]$. A total of 800 samples using Box–Jenkins time series data are used for training and testing the model. The performance of the model is studied, and the error $e(k+1) = N[y_p(k+1)]$ is minimized as shown in Figure 6.30. In Figure 6.31, the outputs of the system as well as the output of FCPN model are shown. Input to the system and the identified model is given by

$$u(k) = \sin(2\pi k/250) \quad \text{for } k \leq 500$$

$$u(k) = 0.8\sin(2\pi k/250) + 0.2\sin(2\pi k/25) \quad \text{for } k > 500$$

Table 6.8 includes various calculated errors of different neural network models for this case. It can be seen from the table that various errors calculated for FCPN are minimum as compared to the DN and backpropagation network (BPN). The BFR for FCPN is maximum; hence, it is concluded that the developed FCPN gives excellent performance. This can also be validated by implementing the same in case of reactive distillation process and described in the next section [21–55].

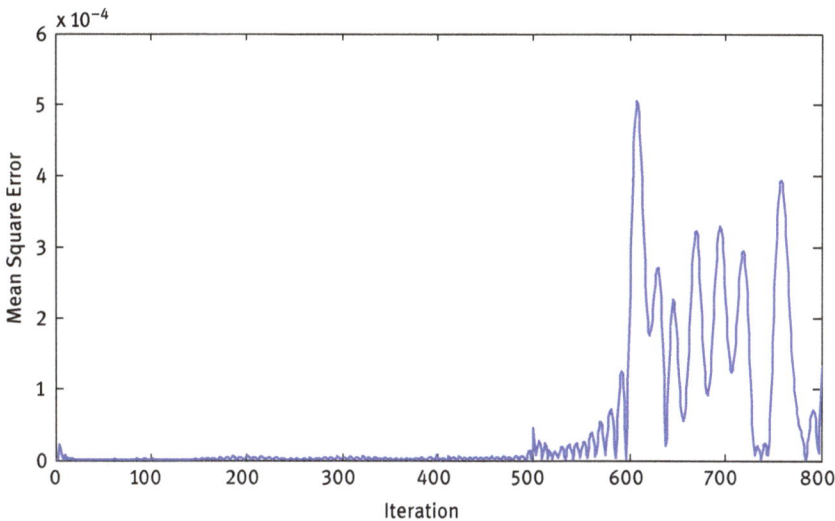

Figure 6.29: Mean square error of the system using FCPN [55].

6.4.2 FCPN for methyl acetate reactive distillation process

After implementing the FCPN for a nonlinear dynamical system, the FCPN is used to implement in case of reactive distillation process. The FCPN controller is designed and implemented to control the sensitive tray temperature profile of the methyl acetate reactive distillation column. The error in temperature (difference between set point temperature and temperature of the tray) and the rate of change of error in temperature are given as input to the FCPN. The change in the feed flow rate is the target output from the FCPN controller. This input–output pattern is used to train the FCPN controller. A total of 500 samples are collected and used for this

Table 6.8: Various errors and BFR for the nonlinear dynamical system [38–55].

NN models	MAE	SSE	MSE	RMSE	NMSE	%BFR
FCPN	30.4348	2.1931	0.0027	0.0524	0.0186	86.35
Dynamic	32.8952	6.4059	0.0080	0.0895	0.0276	83.39
BP	53.1988	13.4174	0.0168	0.1295	0.0844	70.94

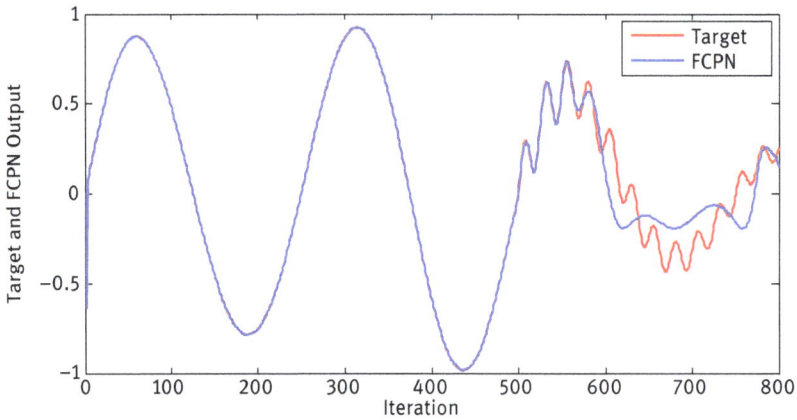

Figure 6.30: Performance of the controller using FCPN algorithm [38–55].

purpose. The performance of the FCPN is compared with DN and BPN. It is observed from the results that FCPN gives superior performance as compared to DN and BPN. The error obtained in case of FCPN is very small as compared to DN and BPN.

The performance of FCPN for the third tray is shown in Figure 6.31 and the performance of FCPN for the sixth tray is shown in Figure 6.32. As shown in figure, the performance of controllers is comparable. However, DN and BPN have oscillatory response as compared to FCPN and later reached to steady state near the set point.

The performance of these controllers is also evaluated by calculating MSE. The plot of MSE for third tray and sixth trays is shown in Figures 6.33 and 6.34, respectively. It is clear from the figure that the error for DN is much large as compared to BPN. In case of FCPN, the error is too small and after 10 s it is almost zero. This shows the superior performance of FCPN controller.

This FCPN controller is compared with DN and BPN for obtaining variations in the distillate composition. The distillate composition using the third tray temperature is shown in Figure 6.35. As per the reaction chemistry, acetic acid and methanol reacted to produce methyl acetate and water. The composition of methyl acetate at the top product is obtained as 95%. It is observed from the figure that FCPN

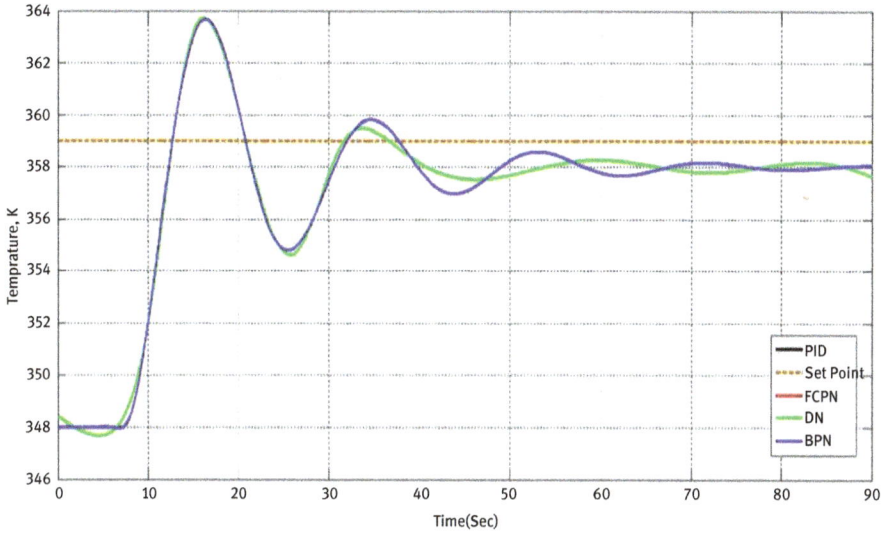

Figure 6.31: Performance of controllers for the third tray temperature [54–55].

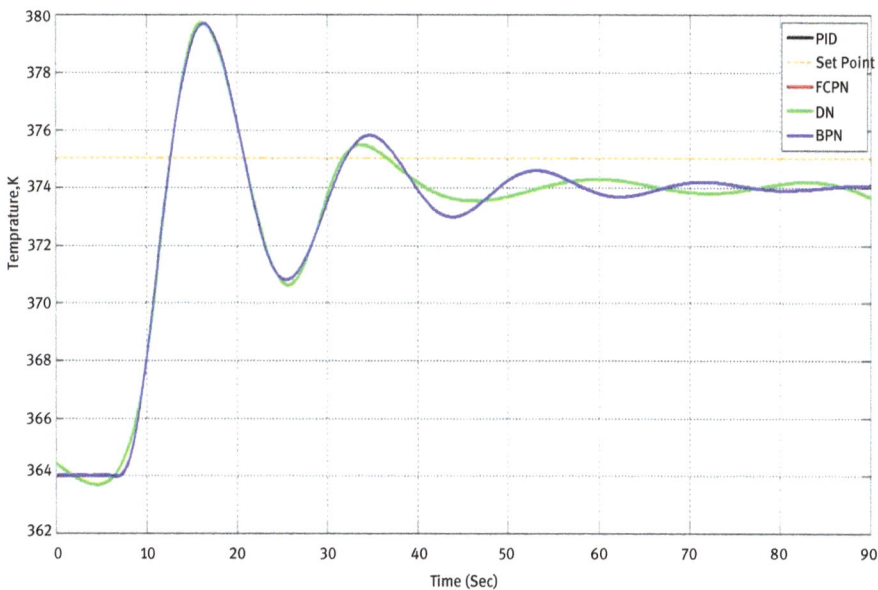

Figure 6.32: Performance of controllers for the sixth tray temperature [54–55].

outperforms over the PID, DN, and BPN [38–55]. The performance of DN and BPN are slightly sluggish. The plots for the error generated are shown in Figure 6.36.

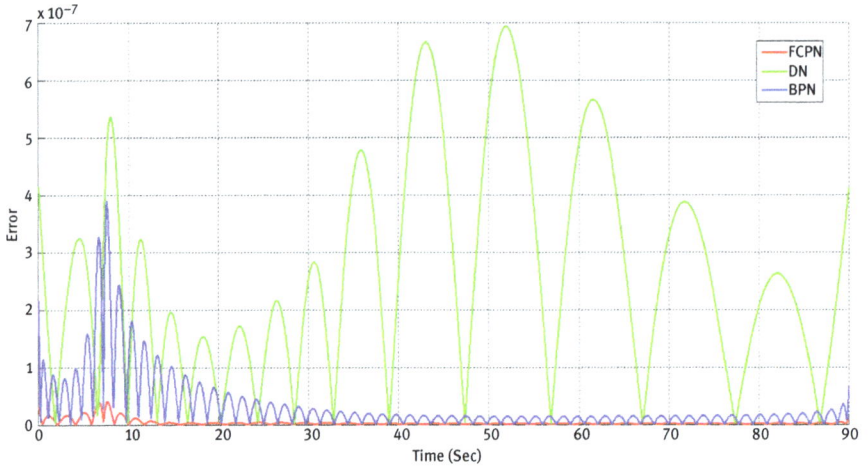

Figure 6.33: Mean square error for the third tray temperature [54–55].

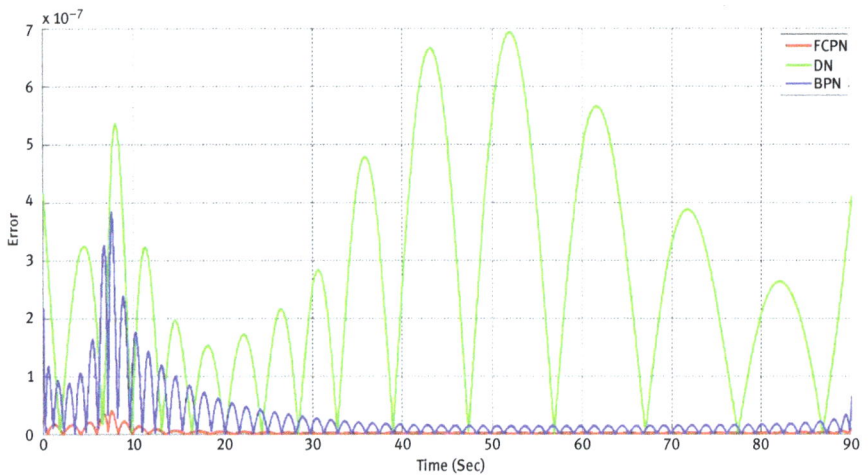

Figure 6.34: Mean square error for the sixth tray temperature [54–55].

It is clear from the figure that FCPN has almost zero error, while BPN has maximum error.

The performance criteria are also evaluated by calculating various errors such as MSE, MAE, SSE, and BFR [38–55]. These errors are given in Table 6.9. It is clear from the table that BFR for FCPN obtained is 99.8429%, which is maximum as compared to DN and BPN. Similarly all these errors are tabulated in Table 6.10 for tray 6.

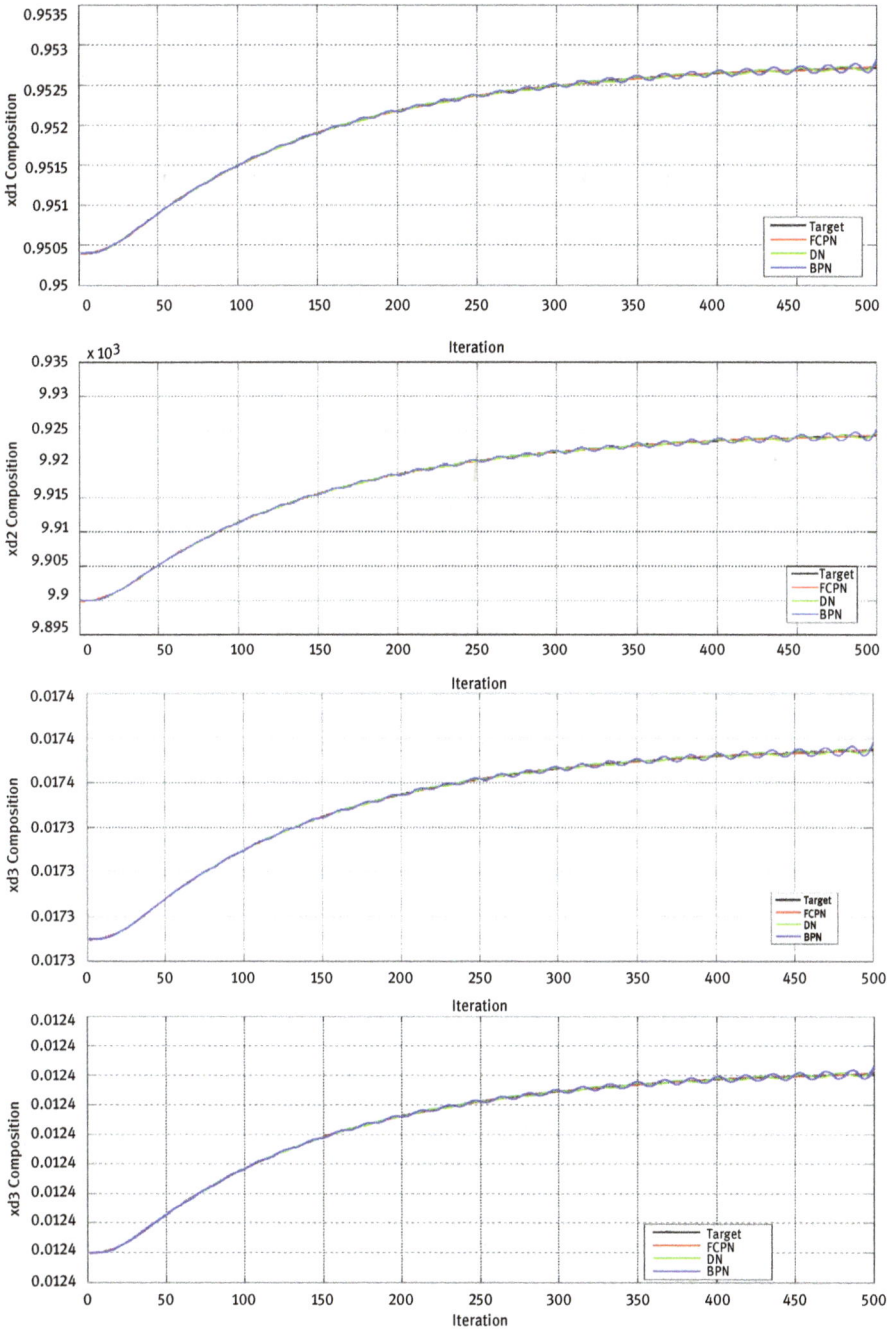

Figure 6.35: Distillate composition using the third tray temperature [53–55].

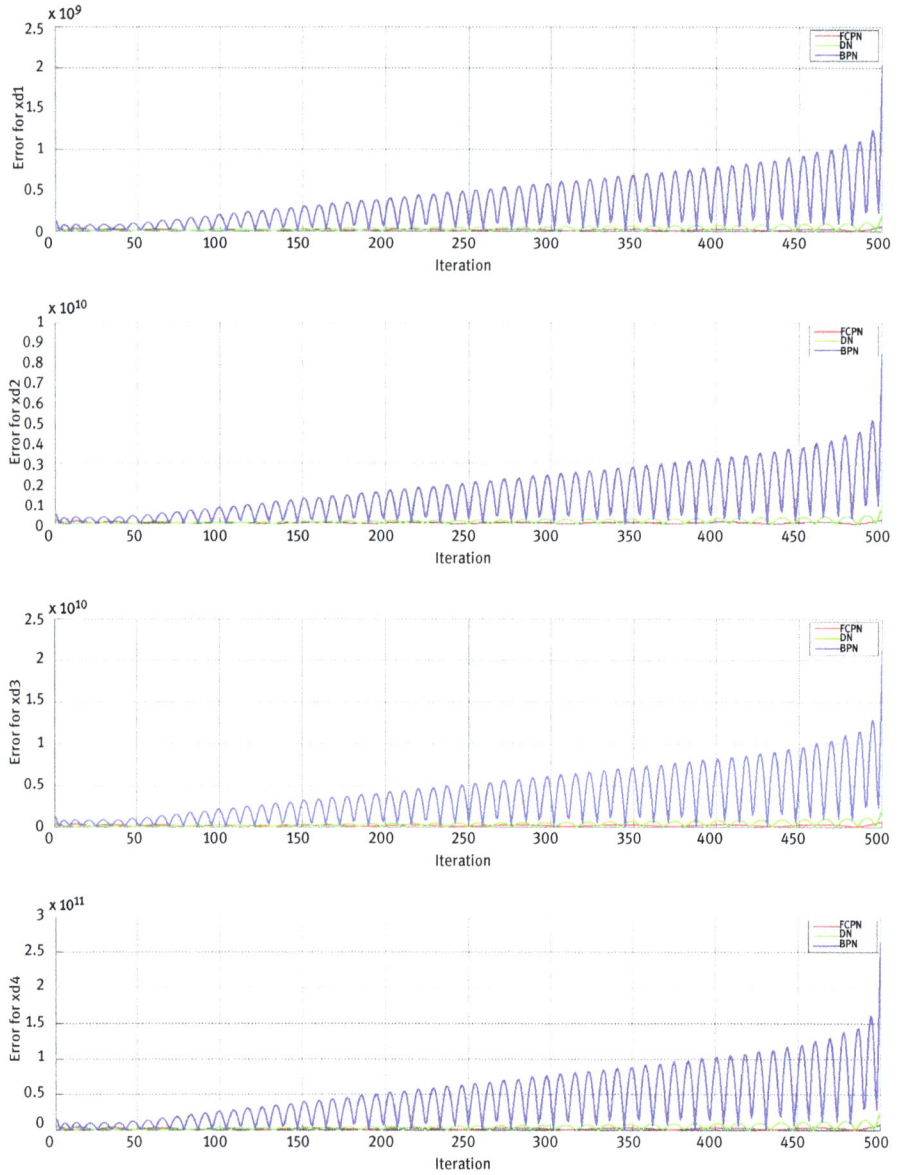

Figure 6.36: Error comparison of controllers for third tray temperature [53–55].

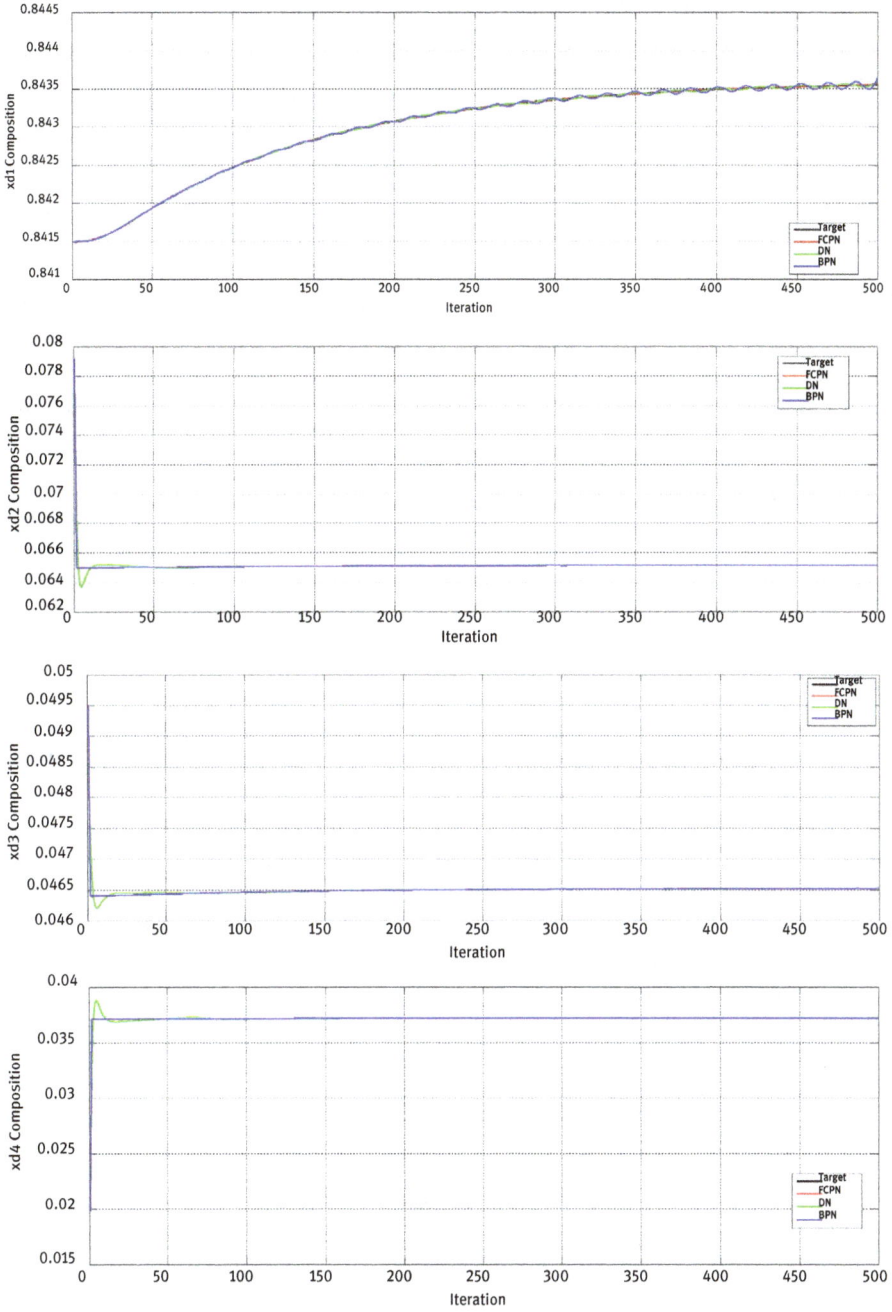

Figure 6.37: Bottom composition using tray 3 temperature [53–55].

Table 6.9: Various errors obtained for FCPN, DN, and BPN for the third tray [29–55].

Neural network	MAE	SSE	MSE	NMSE	BFR (%)
BPN	0.0082	2.3088e-007	4.6177e-010	1.6760e-005	99.5906
DN	0.0037	4.7180e-008	3.5172e-010	3.5679e-006	99.8111
FCPN	0.0018	8.1699e-009	1.1070e-010	2.4673e-006	99.8429

Similarly, the performance of FCPN, DN, and BPN is also obtained and compared for the third tray temperature for the variation in the bottom composition of the methyl acetate reactive distillation column. It is observed that the performance of the FCPN again is superior as compared to DN and BPN. Various errors obtained for FCPN, DN and BPN for the third tray is given in Table 6.9 and various error obtained for FCPN, DN and BPN for the sixth tray is given in Table 6.10.

Similar procedure is followed for the sixth tray temperature. The distillate composition for sixth tray temperature is plotted and shown in Figure 6.38. The maximum distillate composition obtained at top is 96.25%. Here also the performance of FCPN over DN and BPN is superior. The error plot for the same is shown in Figure 6.39. It is clear from the figure that DN has minimum error as compared to FCPN, and BPN has large error.

Similarly, variation in bottom composition for sixth tray temperature is also obtained for the three controllers and plotted. These are shown in Figure 6.40. The maximum bottom composition is obtained as 87.3%. It is clear from the figure that again FCPN performs best as compared to DN and BPN. DN and BPN have slightly sluggish behavior. It is concluded that the target product composition is achieved using FCPN, DN, and BPN for both distillate and bottom composition of methyl acetate reactive distillation. FCPN performs very well as compared to DN and BPN. The various errors obtained for sixth tray temperature are included in Figure 6.39. It is clear from the figure that maximum BFR obtained is for FCPN [29–55].

6.4.3 Experimental validation

The experimental validation of these soft sensors is carried out using pilot-scale reactive distillation studies. The plot of experimental temperature profile is shown in Figure 6.41. The temperature of the reactive zone is increasing due to exothermic reaction and maximum temperature is at reboiler. The composition obtained experimentally is 96%. The comparative plot of experimental and simulation composition is shown in Figure 6.42. As shown in figure, the maximum composition

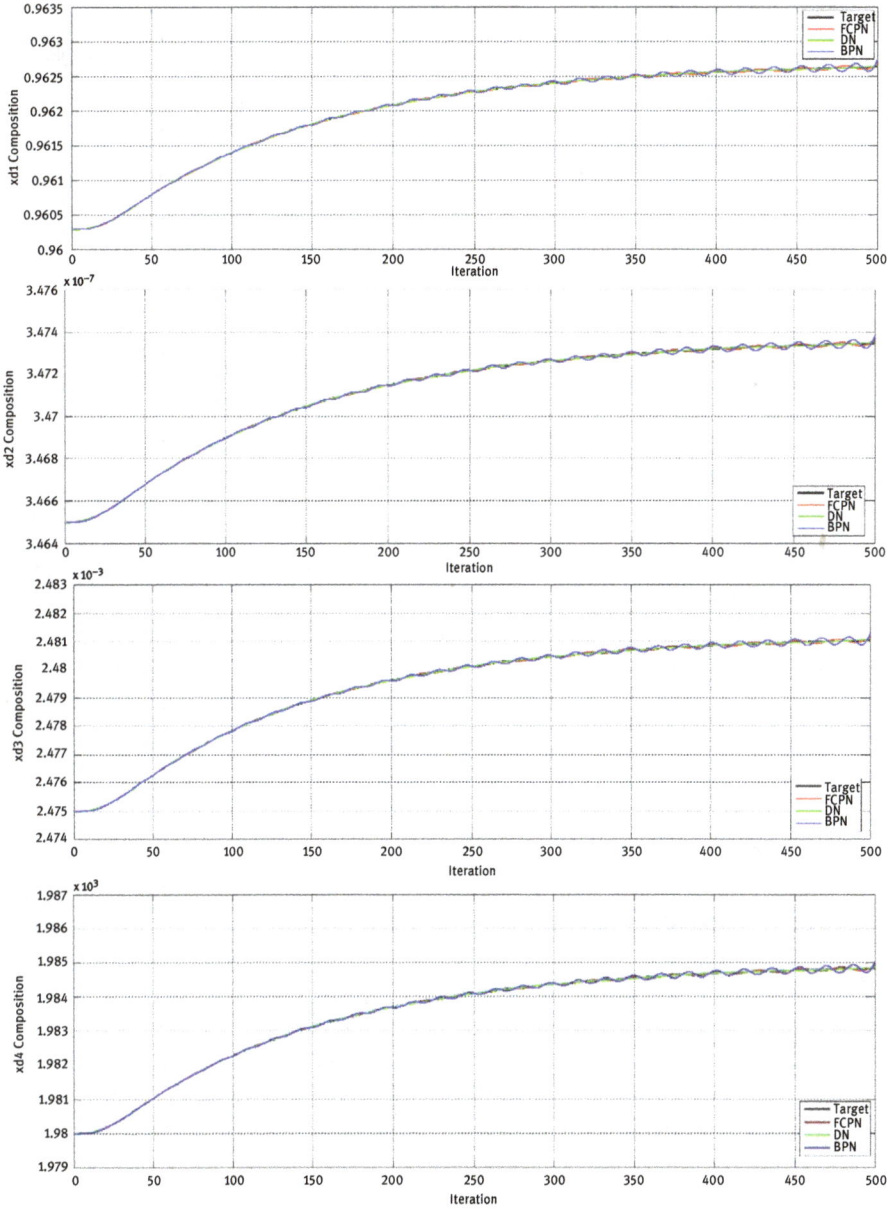

Figure 6.38: Distillate composition using sixth tray temperature [53–55].

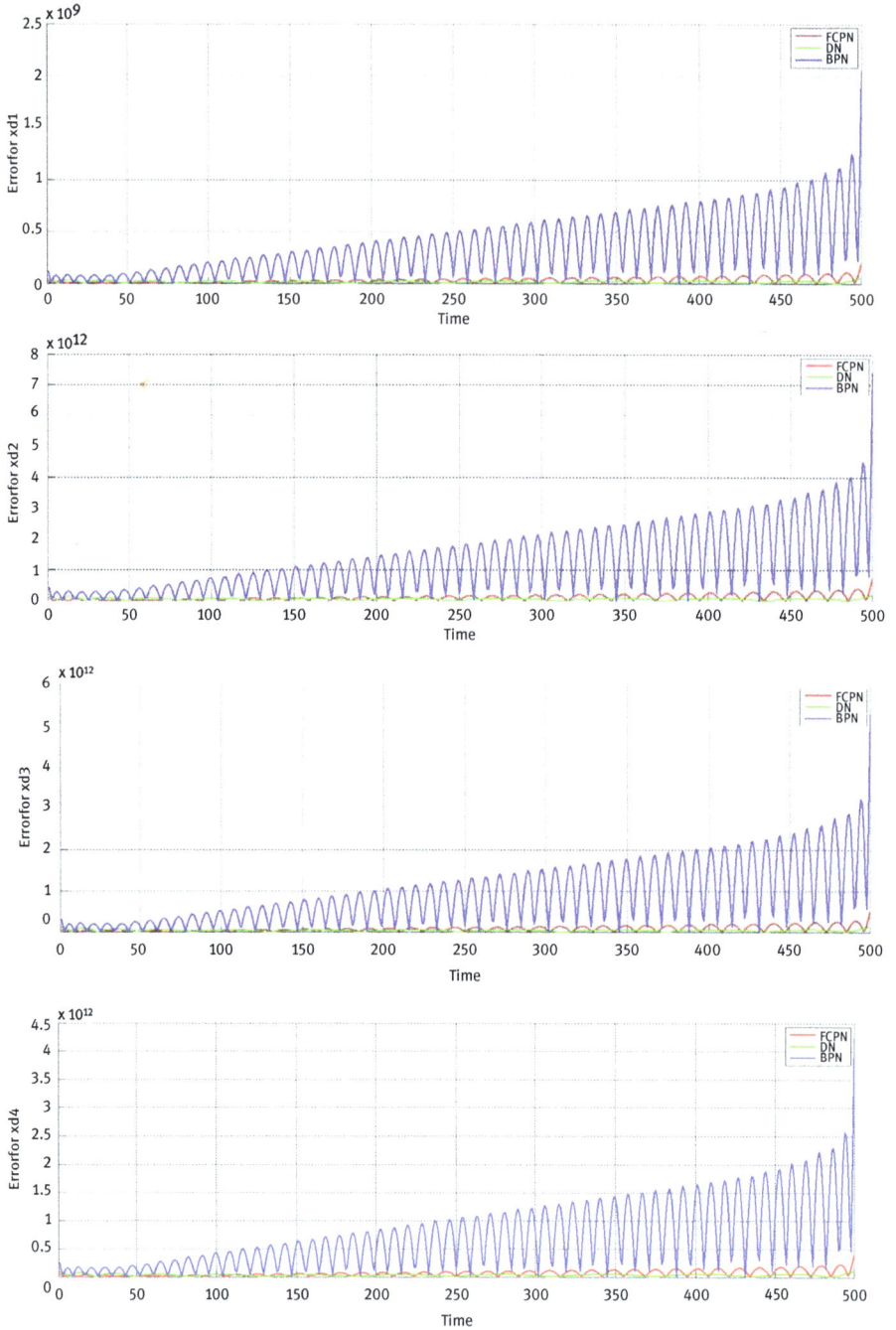

Figure 6.39: Error comparison of controllers for sixth tray temperature [53–55].

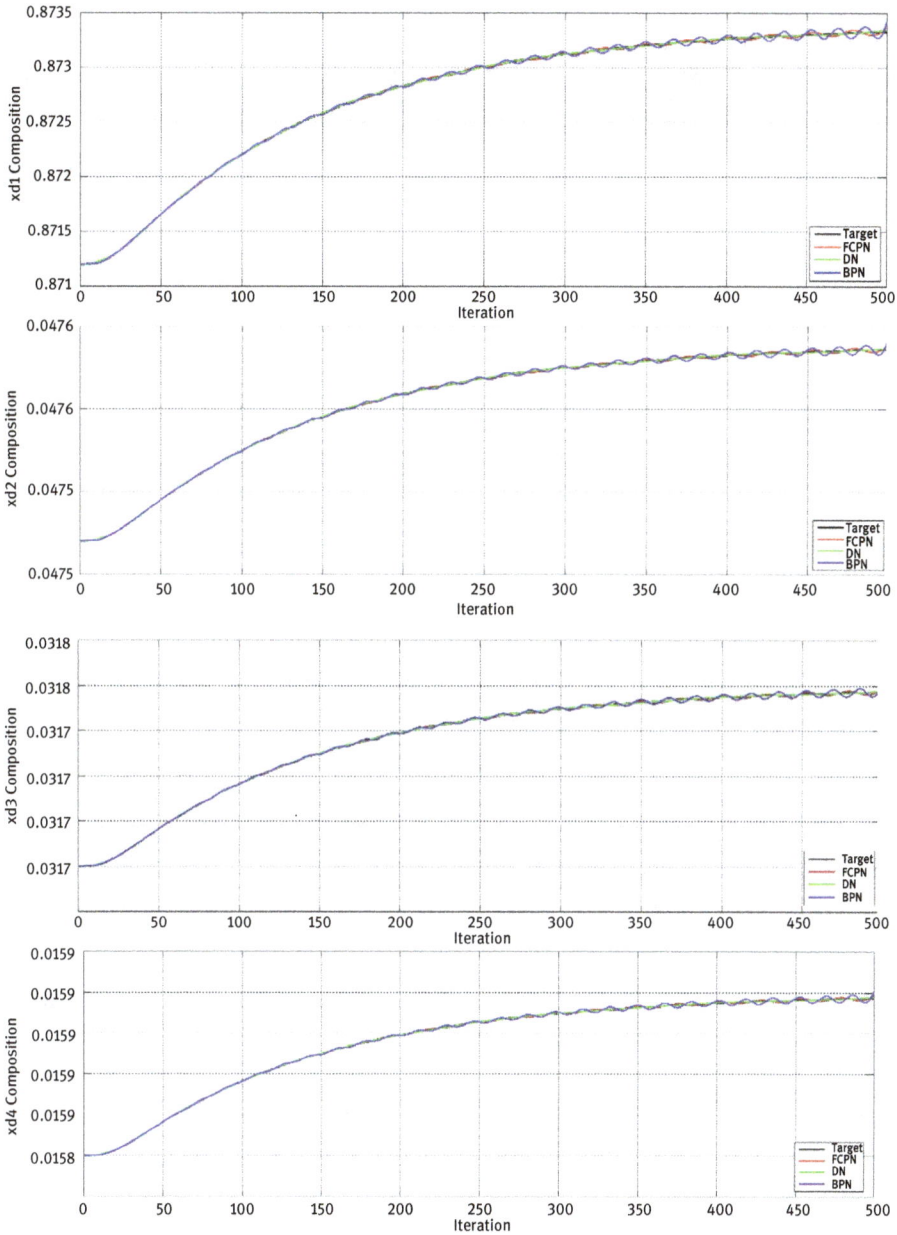

Figure 6.40: Bottom composition using sixth tray temperature [53–55].

Table 6.10: Various errors obtained for FCPN, DN and BPN for 6th tray [27–55].

Neural network	MAE	SSE	MSE	NMSE	BFR (%)
BPN	3.4304e-004	8.1910e-011	8.0168e-013	5.6760e-006	97.9832
DN	1.5360e-004	4.0084e-011	5.0255e-013	3.5679e-006	97.3256
FCPN	7.3300e-005	1.4184e-011	1.9219e-013	2.4673e-006	99.5211

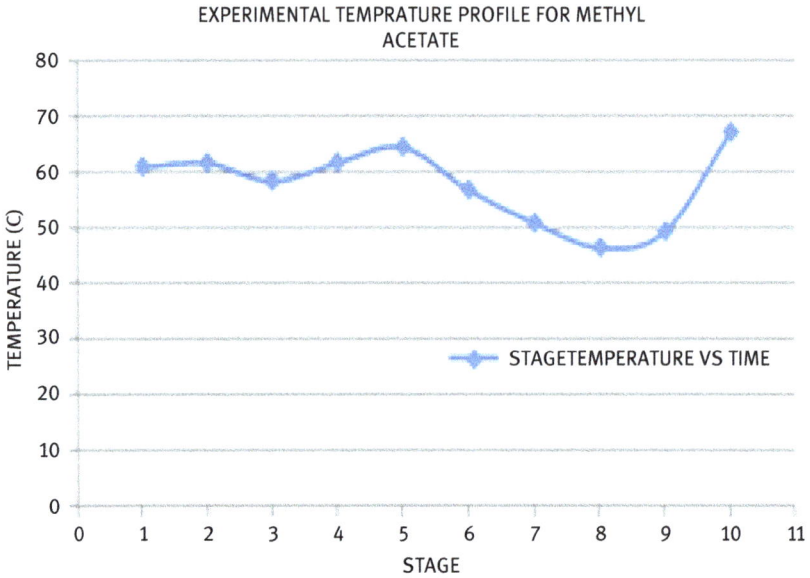

Figure 6.41: Experimental temperature profile [53–55].

obtained experimentally and by simulation is same as 96%. There is a good agreement between experimental and simulation results.

COMPARATIVE GRAPH OF COMPOSITION AS
OBTAINED FROM EXPERIMENT AND SIMULATION

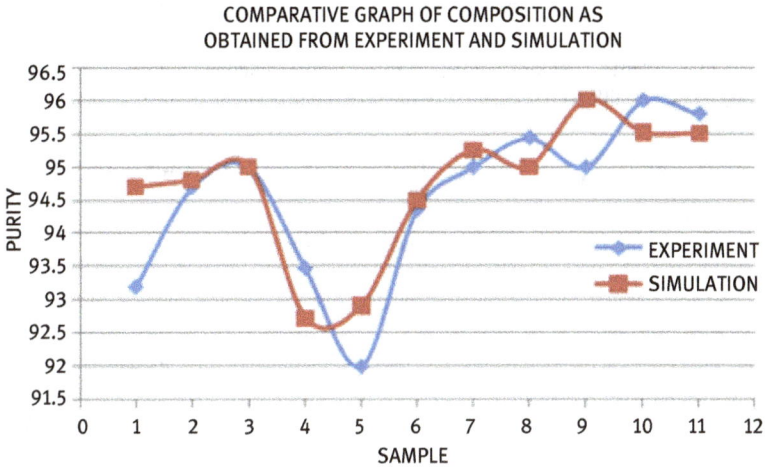

Figure 6.42: Comparative plot of experimental and simulation composition [53–55].

References

[1] P.V.S. Ravi Chandra, C. Venkateshwarlu, Multistep model predictive control of ethyl acetate reactive distillation column, Indian Journal of Chemical Technology, 14, 2007, 333–340.

[2] S. Goyal, A. Rani, V. Singh, SVR tuned PID controller design for reactive distillation process, International Journal of Applied Engineering Research, 7, 2012, 1–6.

[3] M.S. Nizami, Development of a Fuzzy Logic Controller for a Distillation Column Using Rockwell Software, Thesis, 2007.

[4] Z. Lei, C. Yi, B. Yang, Design, optimization, and control of reactive distillation column for the synthesis of tert-amyl ethyl ether, Chemical Engineering Research & Design, 91, 2013, 819–830.

[5] A.C. Dimian, C.S. Bildea, F. Omata, A.A. Kiss, Innovative process for fatty acid esters by dual reactive distillation, Computers & Chemical Engineering, 33, 2009, 743–750.

[6] K. Konakom, A. Saengcham, P. Kittisupakorn, High purity ethyl acetate production with a batch reactive distillation column using dynamic optimization strategy, Proceedings of the World Congress on Engineering and Computer Science, 2, 2010, 978–988.

[7] M.G. Sneesby, O.M. Tade, T.N. Smith, Two-point control of a reactive distillation column for composition and conversion, Journal of Process Control, 9, 1997, 19–31.

[8] F. Zhao, J. Ou, W. Du, Pattern based fuzzy predictive control for a chemical process with dead time, Engineering Applied of Artificial Intelligence, 13, 2000, 37–45.

[9] J.E. Seem, A new pattern recognition adaptive controller with application to HVAC systems, Automatica, 34, 1998, 969–982.

[10] M.J. Jang, C.L. Chen, Fuzzy successive modelling and control for time-delay system, International Journal of Systems Science, 27, 1996, 1483–1490.

[11] Y.C. Tian, Inference of conversion and purity for ETBE reactive distillation, Brazilian Journal of Chemical Engineering, 17, 2000, 617–625.

[12] V. Bansal, V. Sakizlis, R. Ross, J.D. Perkins, New algorithm for mixed integer dynamic optimization, Computers & Chemical Engineering, 21, 2003, 647–668.

[13] S.C. Patwardhan, S. Narasimhan, J. Prakash, R.B. Gopaluni, S.L. Shah, Nonlinear Bayesian state estimation, review and recent trends, Control Engineering Practice, 20, 2012, 933–953.

[14] V.M. Becerra, P.D. Roberts, G.W. Grifiths, Applying the extended Kalman filter to systems described by nonlinear differential-algebraic equations, Control Engineering Practice, 9, 2001, 267–281.

[15] R.K. Mandela, R. Rengaswamy, S. Narasimhan, Recursive state estimation techniques for nonlinear differential algebraic systems, Chemical Engineering Science, 65, 2010, 4548–4556.

[16] B. Huang, R. Kadali, Dynamic Modeling, Predictive Control and Performance Monitoring, Springer publication, 2009, 1–10.

[17] B.M. Akesson, H.T. Toivonen, A neural network model predictive controller, Journal of Process Control, 16, 2006, 937–946.

[18] N. Sharma, Control of Reactive Distillation- A review, International Journal of Chemical Reactor Engineering, 2010, 4–10.

[19] R. Babusk, Fuzzy Systems, Modeling and Identification, IEEE Transactions on Systems, Man, and Cybernetics, 15, 1985, 116–132.

[20] F.B. Rico, J.M. Gozalvez-Zefrilla, J.L. Diez, A. Santafe-Moros, Modeling and control of a continuous distillation tower through fuzzy techniques, Chemical Engineering Research & Design, 89, 2011, 107–115.

[21] M. Pekkanen, A local optimization method for the design of reactive distillation, Computers & Chemical Engineering, 19, 1995, 235–240.

[22] D.O. Araromi, J.O. Emuoyibofarhe, J.A. Sonibare, Fuzzy hybrid modeling of a reactive distillation column for ethyl acetate process, International Journal of Engineering and Technology, 2, 2012, 888–899.

[23] C. Sumana, C. Venkateshwarlu, Optimal selection of sensors for state estimation in a reactive distillation process, Journal of Process Control, 19, 2009, 1024–1035.

[24] X. Wang, R. Luo, H. Shao, Designing a soft sensor for a distillation column with fuzzy distributed radial basis function neural network, Decision and Control, 2, 1996, 1714–1719.

[25] J. Zilkova, Nonlinear System Control Using Neural Networks, Acta Polytechnica, 3, 2006, 85–90.

[26] B.M. Akesson, H.T. Toivonen, A neural network model predictive controller, Journal of Process Control, 16, 2006, 937–946.

[27] S.R. Vijaya Raghavan, T.R. Radhakrishnan, K. Srinivasan, Soft sensor based composition estimation and controller design for an ideal reactive distillation column, ISA Transactions, 50, 2011, 61–70.

[28] K.J. Jithin Prakash, Neuro estimator-based GMC control of batch reactive distillation, ISA Transaction, 50, 2011, 537–539.

[29] A. Bahar, C. Ozgen, State estimation and inferential control for a reactive batch distillation column, Engineering Applications of Artificial Intelligence, 23, 2010, 260–270.

[30] S. Nithya, N. S Ivakumaran, T. Balasubramanian, Controller's implementation based on soft computing for non-linear process, Proceedings of the World Congress on Engineering and Computer Science, Vol. 2, 2010, 978–988.

[31] E. Idris, S. Engell, Real time optimization nonlinear control applied to a continuous reactive distillation process, International Federation of Automatic Control, 18, 2011, 4892–4897.

[32] V. Sujatha, R.C. Panda, Control configuration selection for multi-input multi output processes, Journal of Process Control, 23, 2013, 1567–1574.

[33] B. Subudhi, Nonlinear system identification using memetic differential evolution trained neural networks, Neurocomputing, 74, 2011, 1696–1709.

[34] H.-C. Lu, M.-H. Chang, C.-H. Tsai, Parameter estimation of fuzzy neural network controller based on a modified differential evolution, Neurocomputing, 89, 2012, 178–192.

[35] M. Lawrynczuk, Explicit nonlinear predictive control algorithms with neural approximation, Neurocomputing, 129, 2014, 570–584.

[36] K.K.C.W. Kandanapitiya, Modeling of Reactive Distillation for Acetic Acid Esterification, Journal of the Institution of Engineers, 48, 2015, 17–23.

[37] M. Wierschem, Continuous Enzymatic Reactive Distillation with Immobilized Enzyme Beads, AICHe, Annual meeting, 2015.

[38] T. Zhang, X. Xia, Decentralized adaptive fuzzy output feedback control of stochastic nonlinear large-scale systems with dynamic uncertainties, Information Sciences, 315, 2015, 17–18.

[39] Z. Peng, D. Wang, H. Zhang, Y. Lin, Cooperative output feedback adaptive control of uncertain nonlinear multi-agent systems with a dynamic leader", Neurocomputing, 149, 2015, 132–141.

[40] G. Cui, Z. Wang, G. Zhuang, C. Yuming, Adaptive Centralized NN control of large scale stochastic nonlinear time delay systems with unknown dead zone inputs, Neurocomputing, 158, 2015, 194–203.

[41] Rautenbach, Separation potential of pervaporation, Journal of Membrane Science, 25, 1997, 25–31.

[42] F. Lipnitzki, R.W. Field, P.K. Ten, Pervaporation-based hybrid process: A review of process design, applications and economics, Journal of Membrane Science, 153, 1999, 183–210.

[43] P. Kries, A. Gorak, Process analysis of hybrid separation processes combination of distillation and pervaporation, Chemical Engineering Research and Design, 84, 2006, 595–600.

[44] S. Steinigeweg, J. Gmehling, Transesterification processes by combination of reactive distillation and pervaporation, Chemical Engineering and Processing, 43, 2004, 447–456.

[45] J. Von Scala, E.G. Fassler, Maus, Kontinuierliche Herstellung von kosmetischen Fetts aureestern mittels Reaktivrektifikation and Pervaporation, Chemie Ingenieur Technik., 77, 2005, 1809–1813.

[46] S.S. Ozdemir, Catalytic polymeric membrane: preparation & application, Applied Catalysis A: General, 307, 2006, 167–183.

[47] G.S. Luo, M. Niang, P. Schaetzel, Separation of ethyl tert-butyl ether-ethanol by combined pervaporation and distillation, Chemical Engineering Journal, 68, 1997, 139–143.

[48] A. Gorak, Reactive and membrane assisted distillation: Recent developments and perspective, Chemical Engineering Research & Design, 91, 2013, 1978–1991.

[49] J. Holtbruegge, M. Wierschem, Hybrid configuration of reactive distillation and vapor permeation for the production of dimethyl carbonate and propylene glycol, A Thesis, 2013.

[50] L.V. Bida, G. Liu, X. Dong, W. Wei, W. Jin, Novel reactive distillation–pervaporation coupled process for ethyl acetate production with water removal from reboiler and acetic acid recycle, Industrial and Engineering Chemistry Research, 51, 2012, 8079–8086.

[51] H. Masjuki, Biofuels as diesel fuel alternative: An overview, Journal of Energy Heat Mass Transfer, 15, 1993, 293–304.

[52] V. Sakhre, S. Jain, V.S. Sapkal, D.P. Agarwal, Novel process integration of biodiesel blend in membrane assisted reactive divided wall (MRDW) column, Polish Journal of Chemical Technology, 18, issue 1, 2016, 105–112.

[53] V. Sakhre, S. Jain, V.S. Sapkal, D.P. Agarwal, Modified neural network based cascaded control for product composition of reactive distillation, Polish Journal of Chemical Technology, 18, 2016, 111–121.

[54] V. Sakhre, U.P. Singh, S. Jain, FCPN approach for uncertain nonlinear dynamical system with unknown disturbance, International Journal of Fuzzy System, appeared online 18- 02-2016, Springer, 1–18, 2016.

[55] Vandana Sakhre, Sanjeev Jain, Uday P. Singh, "Fuzzy Induced Counter Propagation Neural Network (FCPN) for the Control of Reactive Distillation Column", Journal of Advanced Research in Dynamics and Control, vol 10 issue 13, 2018, pp 184–192.

Index

https://doi.org/10.1515/9783110656268-007

www.ingramcontent.com/pod-product-compliance
Lightning Source LLC
Chambersburg PA
CBHW081539220326

41598CB00036B/6495